JUST IN TIME Algebra

# Algebra

## Colleen Schultz

NEW YORK

Library of Congress Cataloging-in-Publication Data:
Schultz, Colleen.
    Just in time algebra / Colleen Schultz.—1st ed.
        p. cm.
    ISBN 1-57685-505-8
    1. Algebra—Problems, exercises, etc. I. Title.
QA152. 3. S38 2004
512' .0076—dc22

                                          2003021930

Printed in the United States of America
9 8 7 6 5 4 3 2 1
First Edition

ISBN 1–57685–505–8

For more information or to place an order, contact LearningExpress at:
    55 Broadway
    8th Floor
    New York, NY 10006

Or visit us at:
    www.learnatest.com

# ABOUT THE AUTHOR

**Colleen Schultz** is a mathematics teacher/teacher mentor for the Vestal School District in Vestal, NY. She is the math author of *GMAT Exam Success* and is a contributing writer for *501 Math Word Problems* and *501 Quantitative Comparison Questions*. Colleen is also an expert math reviewer and math item writer for several high-stakes math practice tests. In addition, she is a teacher trainer for the use of technology in the mathematics classroom.

# CONTENTS

# Introduction

In just a few short weeks—or maybe just a few days—you will be taking that important exam that will test your algebra skills and you have yet to begin to study. Perhaps time is a factor for you; your schedule is packed with work, family, and other duties that keep you from preparing for the exam. Or possibly you have had the time, but you have procrastinated; algebra has never been your best subject. Maybe you have just put off studying since you only need to brush-up on skills, rather than complete a comprehensive review course. Or maybe you have just realized that this exam includes a mathematical section that will test algebra skills, and now there is only a limited amount of time to get ready.

If you find yourself fitting into any of these scenarios, then *Just in Time Algebra* is just the book for you. Designed especially for last-minute exam preparation, *Just in Time Algebra* is a fast, accurate method to review and practice those essential algebra skills. With over 300 algebra questions that address commonly tested concepts, this workbook will help you review the knowledge and skills you have already mastered and teach you other concepts and strategies that you will need for the test. In just ten short chapters, you will get just the essentials, just in time for passing your big test.

## THE *JUST IN TIME* TEST-PREP APPROACH

At LearningExpress, we know how important test scores and a solid foundation of algebra skills can be. Whether you are preparing for a high-stakes test, such as the PSAT® tests, SAT® exam, GRE® test, GMAT®, a civil service exam, or you simply need to improve your fundamental algebra skills fast, the *Just in Time* streamlined approach can work for you. Each skill-building lesson includes:

- a total of 35 algebra problems incorporating commonly tested skills
- a ten-question benchmark quiz to help you assess your knowledge of the concepts and skills in the chapter
- a brief, focused lesson covering essential algebraic topics, skills, and applications
- specific tips and strategies to use as you study and take the exam
- a 25-question practice quiz followed by complete answer explanations to help you assess your progress

The *Just in Time* series also includes the following features:

-  *Extra Help:* refers you to other LearningExpress skill builders or other resources that can help you learn more about a particular topic.

-  *Calculator Tips:* hints on how your calculator can help you.

- ⬤⬤⬤⬤⬤⬤ *Glossary:* provides critical definitions.

- *Rule Book:* highlights the rules and procedures you really need to know.

- *Shortcut:* suggests tips for reducing your study and practice time—without sacrificing accuracy.

- *Formula Cheat Sheet:* a tear-out page following this introduction that lists important formulas from the chapters.

As you probably realize, no algebra book can possibly cover all of the skills and concepts you may be faced with on a standardized test, and here we have limited our questions to just over 300. However, this book is not just about building an algebra base; it is also about building those essential skills that can help you solve questions that you did not already know how to do. The algebra topics and skills in this book have been carefully selected to represent a cross-section of basic skills that can be applied in a more complex setting, as needed.

## HOW TO USE THIS BOOK

The chapters in this book cover general study skills along with a broad spectrum of algebra topics. Chapter 1 covers practical, last-minute study skills, and Chapters 2–10 cover specific algebraic concepts common to most standardized tests. While each chapter can stand on its own as an effective algebraic skill builder, this book will be most effective if you complete each chapter in order, beginning with Chapter 1. Algebra is a sequential and cumulative subject in which more complex skills build on a previous foundation of concepts. The chapters increase slightly in difficulty as the book progresses, so you can master the easier concepts first before taking on the more complicated topics.

Here is an outline of each chapter:

- **Chapter 1: Study Skills** reviews fundamental study strategies, including how to budget your time, create a study plan, and use study aids, such as flashcards.
- **Chapter 2: Number Properties and Equation Solving** covers the properties of real numbers, order of operations, and how to solve various types of equations.
- **Chapter 3: Coordinate Geometry and Graphing Linear Equations** covers how to graph points and lines in the coordinate plane.
- **Chapter 4: Systems of Equations** covers how to solve systems of equations both graphically and algebraically.
- **Chapter 5: Linear and Compound Inequalities** covers how to solve and graph inequalities of one and two variables, as well as compound inequalities.
- **Chapter 6: Exponents** covers how to simplify and perform operations with exponents and radicals.
- **Chapter 7: Polynomials** covers how to simplify and perform operations with polynomials.
- **Chapter 8: Factoring and Quadratic Equations** covers factoring various types of polynomials and how to solve quadratic equations.
- **Chapter 9: Algebraic Fractions** covers how to simplify and perform operations with rational expressions and how to solve rational equations.
- **Chapter 10: Translating Algebraic Expressions and Solving Word Problems** covers how to solve different types of common

algebraic word problems, as well as how to translate words into algebraic expressions and equations.

Depending on the amount of time you have before the exam, review as much of the material as possible. Brush up on the skills and concepts from each chapter you have completed before you move on to the next. That way, you will continue to reinforce your knowledge of the skills you have already covered before you add more to your algebraic repertoire.

You can do it! That important exam may be just around the corner, but you are taking the critical steps to get ready . . . *just in time*.

# FORMULA CHEAT SHEET

## ORDER OF OPERATIONS

Please Excuse My Dear Aunt Sally
1. **P** **P**arentheses/grouping symbols (including fraction bars) first
2. **E** then **E**xponents
3. **MD** **M**ultiplication/**D**ivision in order from right to left
4. **AS** **A**ddition/**S**ubtraction in order from right to left

## PROPERTIES OF REAL NUMBERS

Commutative Property
$$a + b = b + a \qquad ab = ba$$
Associative Property
$$a + (b + c) = (a + b) + c \qquad a(bc) = (ab)c$$
Identity Property
$$a + 0 = a \qquad a \cdot 1 = a$$
Inverse Property
$$a + (-a) = 0 \qquad a \cdot \frac{1}{a} = 1$$
Distributive Property
$$a(b + c) = ab + ac \quad \text{or} \qquad a(b - c) = ab - ac$$

## COORDINATE GRAPHING

Slope-Intercept Form of a Linear Equation
$y = mx + b$, where $m$ is the slope of the line and $b$ is the $y$-intercept

Midpoint Formula
$$\left(\frac{x_1 + x_2}{2}, \frac{y_1 + y_2}{2}\right)$$

Slope Formula
$$m = \frac{\text{change in } y}{\text{change in } x} = \frac{y_1 - y_2}{x_1 - x_2}$$

Distance Formula
$$d = \sqrt{(x_1 - x_2)^2 + (y_1 - y_2)^2}$$

## EXPONENTS

### Multiplying Like Bases
$$x^a \cdot x^b = x^{a+b}$$

### Dividing Like Bases
$$\frac{x^a}{x^b} = x^{a-b}$$

### Power Raised to a Power
$$(x^a)^b = x^{a \cdot b}$$

### Product to a Power
$$(xy)^a = x^a y^a$$

### Quotient to a Power
$$\left(\frac{x}{y}\right)^a = \frac{x^a}{y^a}$$

### Exponent of Zero
$$x^0 = 1$$

### Negative Exponents
$$x^{-a} = \frac{1}{x^a}$$

## RADICALS

### Roots of Expressions
$$\sqrt{x} = x^{\frac{1}{2}}$$
$$\sqrt[a]{x} = x^{\frac{1}{a}}$$

### Multiplication
$$\sqrt{x} \cdot \sqrt{y} = \sqrt{xy}$$

### Division
$$\sqrt{\frac{x}{y}} = \frac{\sqrt{x}}{\sqrt{y}}$$

## POLYNOMIALS, FACTORING, AND QUADRATICS

### The Difference Between Two Squares
$$x^2 - y^2 = (x + y)(x - y)$$

### Squaring Binomials
$$(x + y)^2 = (x + y)(x + y) = x^2 + 2xy + y^2$$
$$(x - y)^2 = (x - y)(x - y) = x^2 - 2xy + y^2$$

### Standard Form of a Quadratic Equation
$$ax^2 + bx + c = 0$$

### The Quadratic Formula

$$x = \frac{-b \pm \sqrt{b^2 - 4ac}}{2a}$$

# Study Skills

If you have left studying for that big test until the last minute, you may be feeling that your only option is to cram. You might be feeling panicky that you will never have enough time to learn what you need to know. But the "Just in Time" solution is exactly that: "just in time." This means that with the help of this book you can use your available time prior to your test effectively. First, to get ready for your test "just in time," you need a plan. This chapter will help you put together a study plan that maximizes your time and tailors your learning strategy to your needs and goals.

There are four main factors that you need to consider when creating your study plan: what to study, where to study, when to study, and how to study. When you put these four factors together, you can create a specific plan that will allow you to accomplish more—in less time. If you have three, two weeks or even one week to get ready, you can create a plan that avoids anxiety-inducing cramming and focuses on real learning by following the simple steps in this chapter.

## WHAT TO STUDY

Finding out what you need to study for your test is the first step in creating an effective study plan. You need to have a good measure of your ability in algebra. You can accomplish this by looking over the Table of Contents to see what looks familiar to you and by answering the Benchmark Quiz questions starting in the next chapter. You also need to know what exactly is covered on the test you will be taking. Considering both your ability and the test content will tell you what you need to study.

## ▶ *Establish a Benchmark*

In each chapter you will take a short, ten-question Benchmark Quiz that will help you assess your skills. This may be one of the most important steps in creating your study plan. Because you have limited time, you need to be very efficient in your studies. Once you take a chapter Benchmark Quiz and analyze the results, you will be able to avoid studying the material you already know. This will allow you to focus on those areas that need the most attention.

A Benchmark Quiz is only practice. If you did not do as well as you anticipated you might, do not be alarmed and certainly do not despair. The purpose of the quiz is to help you focus your efforts so that you can *improve*. It is important to carefully analyze your results. Look beyond your score, and consider *why* you answered some questions incorrectly. Some questions to ask yourself when you review your wrong answers:

- Did you get the question wrong because the material was totally unfamiliar?
- Was the material familiar but were you unable to come up with the right answer? In this case, when you read the right answer it will often make perfect sense. You might even think, "I knew that!"
- Did you answer incorrectly because you read the question carelessly?
- Did you make another careless mistake? For example, circling choice **a** when you meant to circle choice **b**.
    Next, look at the questions you got right and review how you came up with the right answer. Not all correct answers are created equal.
- Did you simply know the right answer?
- Did you make an educated guess? An educated guess might indicate that you have some familiarity with the subject, but you probably need at least a quick review.
- Or did you make a lucky guess? A lucky guess means that you don't know the material and you will need to learn it.

Your performance on each Benchmark Quiz will tell you several important things. First, it will tell you how much you need to study. For example, if you got eight out of ten questions right (not counting lucky guesses!), you might only need to brush up on certain areas of knowledge. But if you got five out of ten questions wrong, you will need a thorough review. Second, it can tell you what you know well, that is which subjects you *don't* need to study. Third, you will determine which subjects you need to study in-depth and which subjects you simply need to refresh your knowledge.

## ▶ Targeting Your Test

For the "Just in Time" test-taker, it is important to focus your study efforts to match what is needed for your test. You don't want to waste your time learning something that will not be covered on your test. There are three important aspects that you should know about your test before developing your study plan:

- What material is covered?
- What is the format of the test? Is it multiple choice? Fill in the blank? Some combination? Or something else?
- What is the level of difficulty?

How can you learn about the test before you take it? For most standardized tests, there are sample tests available. These tests—which have been created to match the test that you will take—are probably the best way to learn what will be covered. If your test is non-standardized, you should ask your instructor specific questions about the upcoming test.

You should also know how your score will affect your goal. For example, if you are taking the SAT I exam, and the median math score of students accepted at your college of choice is 550, then you should set your sights on achieving a score of 550 or better. Or, if you are taking the New York City Police Officer exam, you know that you need to get a perfect or near-perfect score to get a top slot on the list. Conversely, some exams are simply pass or fail. In this case, you can focus your efforts on achieving a passing score.

## ▶ Matching Your Abilities to Your Test

Now you understand your strengths and weaknesses and you know what to expect of your test, you need to consider both factors to determine what material you need to study. First, look at the subject area or question type with which you have the most trouble. If you can expect to find questions of this type on your test then this subject might be your first priority. But be sure to consider how much of the test will cover this material. For example, if there will only be a few questions out of a hundred that test your knowledge of a subject that is your weakest area, you might decide not to study this subject area at all. You might be better served by concentrating on solidifying your grasp of the main material covered on the exam.

The important thing to remember is that you want to maximize your time. You don't want to study material that you already know. And you don't want to study material that you don't need to know. You will make the best use of your time if you study the material that you know the least but that you most need to know.

## WHERE TO STUDY

The environment in which you choose to study can have a dramatic impact on how successful your studying is. If you chose to study in a noisy coffee shop at a small table with dim lighting, it might take you two hours to cover the same material you could read in an hour in the quiet of the library. That is an hour that you don't have to lose! However, for some people the noisy coffee shop is the ideal environment. You need to determine what type of study environment works for you.

### ▶ *Consider Your Options*

Your goal is to find a comfortable, secure place that is free from distractions. The place should also be convenient and conform to your schedule. For example, the library might be ideal in many respects. However, if it takes you an hour to get there and it closes soon after you arrive you are not maximizing your study time.

For many people studying at home is a good solution. Home is always open and you don't waste any time getting there, but it can have drawbacks. If you are trying to fit studying in between family obligations, you might find that working from home offers too many opportunities for distraction. Chores that have piled up, children or younger siblings who need your attention, or television that captures your interest, are just some of things that might interfere with studying at home. Or maybe you have roommates who will draw your attention away from your studies. Studying at home is a good solution if you have a room that you can work in alone and away from any distractions.

If home is not a good environment for quiet study, the library, a reading room, or a coffee shop are places you can consider. Be sure to pick a place that is relatively quiet and which provides enough workspace for your needs.

## ▶ Noise

Everyone has his or her own tolerance for noise. Some people need absolute silence to concentrate, while others will be distracted without some sort of background noise. Classical music can be soothing and might help you relax as you study. If you think you work better with music or the television on, you should be sure that you are not paying attention to what's on in the "background." Try reading a chapter or doing some problems in silence, then try the same amount of work with noise. Which noise level allowed you to work the fastest?

## ▶ Light

You will need to have enough light to read comfortably. Light that is too dim will strain your eyes and make you drowsy. Too bright light may make you uncomfortable and tense. Experts suggest that the best light for reading comes from behind and falls over your shoulder. Make sure your light source falls on your book and does not shine in your eyes.

## ▶ Comfort

Your study place should be comfortable and conducive to work. While your bed might be comfortable, studying in bed is probably more conducive to sleep than concentrated learning. You will need a comfortable chair that offers good back support and a work surface—a desk or table—that gives you enough space for your books and other supplies. Ideally, the temperature should be a happy medium between too warm and too cold. A stuffy room will make you sleepy and a cold room is simply uncomfortable. If you are studying outside your home, you may not be able to control the temperature, but you can dress appropriately. For example, bring along an extra sweater if your local library is skimpy with the heat.

## ▶ A Little Help

When you have settled on a place to study, you will need to enlist the help of your family and friends—especially if you are working at home. Be sure they know that when you go to your room and close the door to study that you do want to be disturbed. If your friends all go to the same coffee shop where you plan to study, you will also need to ask them to respect your study place. The cooperation of your family and friends will eliminate one of the greatest potential distractions.

## WHEN TO STUDY

Finding the time in your busy schedule may seem like the greatest hurdle in making your "just in time" study plan, but you probably have more time available than you think. It just takes a little planning and some creativity.

### ▶ Analyze Your Schedule

Your first step in finding time to study is to map out your day-to-day schedule—*in detail*. Mark a piece of paper in fifteen-minute intervals from the time you get up to the time you generally go to bed. Fill in each fifteen-minute interval. For example, if you work from nine to five, do not simply block that time off as unavailable for study. Write down your daily routine at work and see when you might have some time to study. Lunch is an obvious time. But there may be other down times in your workday when you can squeeze in a short study session.

You will want to set aside a stretch of time when you plan to study in your designated study place. But you can also be creative and find ways to study for short bursts during your normal routine. For example, if you spend an hour at the gym on the stationary bike, you can read while you cycle. Or you can review flashcards on your bus ride. If you drive to work, you could record some study material on a tape or CD. You could also listen to this tape while you walk the dog.

When you look at your schedule closely, you will probably find you have more time than you thought. However, if you still don't have the time you need, you should rethink your routine. Can you ask your significant other to take on a greater share of the household chores for the few weeks you need to get ready for your test? Is there some activity that you can forgo for the next few weeks? If you normally go to the gym six days a week for an hour and a half, cut down to three days a week for forty-five minutes. You will add over six and a half hours to your schedule without completely abandoning your fitness routine. Remember any changes you make to your schedule are short-term and a small sacrifice, once you consider your goal.

### ▶ Time Strategies

Now that you know when you have time available, you need to use that time to your best advantage. You will probably find that you can set aside one block of time during the day during which you will do the bulk of your studying. Use this time to learn new material, or take a practice quiz and review your answers. Use the small spurts of time you have found in your schedule to review with flashcards, cheat sheets, and other tools.

Also consider your learning style and body rhythm when you make your schedule. Does it take you some time to get into material? If so, you should build a schedule with longer blocks of time. Do you have a short attention span? Then you will do better with a schedule of several shorter study periods. No matter your style, avoid extremes. Neither very long study sessions nor very short (except for quick reviews) sessions are an efficient use of time. Whether you are a morning person or a night owl, plan to study when you are most energetic and alert.

Make sure your schedule allows for adequate rest and study breaks. Skipping sleep is not a good way to find time in your schedule. Not only will you be tired when you study, you will be sleep deprived by the time of the test. A sleep-deprived test-taker is more likely to make careless mistakes, lose energy and focus, and become stressed-out by the testing environment. If you plan to do most of your studying in one block of time, say four hours, be sure you leave time to take a study break. Experts have shown that students are more likely to retain material if they take some time to digest it. A five- or ten-minute break to stretch your legs or eat a snack will revive you and give your brain time to absorb what you have learned.

## HOW TO STUDY

How you study is just as important as how long—especially if your time is limited. You will need to be in a good physical and mental state and you will need to use the right tools for the job. You will also need to understand your learning style so that you can select the best study method. And, perhaps most important, you will need methods that will help you to remember, not to memorize, the material. All these techniques—using the right tools and methods—will help you make the most of your study time.

### ▶ *Sleep Well, Eat Right, and Relax*

Does your idea of studying hard include images of staying up into the wee hours and living on fast food and caffeine until the big test? Even though it may seem like you are working hard when you study around the clock and put aside good eating habits in order to save time, you are not working efficiently. If you have ever pulled an "all-nighter" you know that by four in the morning you can find yourself reading the same page several times without understanding a word. Adequate rest and good nutrition will allow you to be focused and energetic so you can get more work done in less time.

Most people need about eight hours of sleep a night. Do not sacrifice sleep in order to make time to study. Hunger can be a distraction, so don't skip meals. Eat three nutritious meals a day, and keep healthy snacks on hand during a long study session. The keyword is *healthy*. Sugary snacks might make you feel energized in the short term, but that sugar rush is followed by a crash that will leave you feeling depleted. Caffeine can have a similar effect. A little caffeine—a morning cup of coffee, for example—can give you a boost, but too much caffeine will make you feel jittery and tense. Tension can affect your ability to concentrate.

Being over-caffeinated is not the only potential source of tension. Pre-exam anxiety can also get in the way of effective studying. If your anxiety about the upcoming test is getting the better of you, try these simple relaxation techniques:

- **Breathe!** Sounds simple, and it is. Taking long, deep breaths can drain the tension from your body. Place one hand on your stomach and the other on your chest. Sit up straight. Inhale deeply through your nose and feel your stomach inflate. Your chest should remain still. Exhale slowly through your mouth and feel your stomach deflate. It is the slow exhalation that helps you relax, so make sure you take your time releasing your breath. Pausing during a study session to take three deep breaths is a quick way to clear your mind and body of tension so that you can better focus on your work.
- **Tense and relax your muscles.** You may not even notice it, but as anxiety mounts your muscles tense. You may tense your neck and shoulders, your toes, or your jaw. This tension can interfere with your concentration. Release the tension held in your muscles by purposefully tensing then relaxing each muscle. Work from your toes to your head systematically.
- **Visualize a soothing place.** Taking a break to mentally visit a place that you find relaxing can be reinvigorating. Close your eyes and conjure up the sights, smells, and sounds of your favorite place. Really try to feel like you are there for five uninterrupted minutes and you will return from your mini vacation ready to study.

## ▶ The Right Tools for the Job

If you follow the steps above, you will have a rested, energized, and relaxed brain—the most important tool you need to prepare for your exam. But there are other tools that you will need to make your study session the most productive. Be sure that you have all the supplies you need on hand before you sit down to study. To help make studying more pleasant, select supplies that you enjoy using. Here is a list of supplies that you will need:

- a notebook or legal pad dedicated to studying for your test
- graph paper
- pencils

- pencil sharpener
- highlighter
- index or other note cards
- paper clips or sticky note pads for marking pages
- a calendar or personal digital assistant (which you will use to keep track of your study plan)
- a calculator

## ▶ *Break It Down*

You may be feeling overwhelmed by the amount of material you have to cover in a short time. This seeming mountain of work can generate anxiety and even cause you to procrastinate further. Breaking down the work into manageable chunks will help you plan your studying and motivate you to get started. It will also help you organize the material in your mind. When you begin to see the large topic as smaller units of information that are connected, you will develop a deeper understanding of the subject. You will also use these small "chunks" of information to build your study plan. This will give you specific tasks to accomplish each day, rather than simply having time set aside to "study for the test."

For example, if you have difficulty factoring equations, you could study a different factoring method each day for a week: On Monday, practice with problems that use the difference of two squares; on Tuesday, work on using the quadratic formula to solve quadratic equations; on Wednesday, try factoring polynomials by grouping; and so on. "Learn to factor equations" might seem like an overwhelming task, but if you divide the work into smaller pieces, you will find that your understanding of factoring improves with practice and patience.

## ▶ *Your Learning Style*

Learning is not the same for everyone. People absorb information in different ways. Understanding how you learn will help you develop the most effective study plan for your learning style. Experts have identified three main types of learners: visual, auditory, and kinesthetic. Most people use a combination of all three learning styles, but one style might be more dominant. Here are some questions that will help you identify your dominant learning style:

1. If you have to remember an unusual word, you most likely
   **a.** picture the word in your mind.
   **b.** repeat the word aloud several times.
   **c.** trace out the letters with your finger.

2. When you meet new people, you remember them mostly by
   a. their actions and mannerisms.
   b. their names (faces are hard to remember).
   c. their faces (names are hard to remember).

3. In class you like to
   a. take notes, even if you don't reread them.
   b. listen intently to every word.
   c. sit up close and watch the instructor.

A visual learner would answer **a**, **c**, and **c**. An auditory learner would answer **b**, **b**, and **b**. A kinesthetic learner would answer **c**, **a**, and **a**.

Visual learners like to read and are often good spellers. When visual learners study, they often benefit from graphic organizers such as charts and graphs. Flashcards often appeal to them and help them learn, especially if they use colored markers, which will help them form images in their minds as they learn words or concepts.

Auditory learners, by contrast, like oral directions and may find written materials confusing or boring. They often talk to themselves and may even whisper aloud when they read. They also like being read aloud to. Auditory learners will benefit from saying things aloud as they study and by making tapes for themselves and listening to them later. Oral repetition is also an important study tool. Making up rhymes or other oral mnemonic devices will also help them study, and they may like to listen to music as they work.

Kinesthetic learners like to stay on the move. They often find it difficult to sit still for a long time and will often tap their feet and gesticulate a lot while speaking. They tend to learn best by doing rather than observing. Kinesthetic learners may want to walk around as they practice what they are learning because using their bodies helps them remember things. Taking notes is an important way of reinforcing knowledge for the kinesthetic learner, as is making flashcards.

It is important to remember that most people learn in a mixture of styles, although they may have a distinct preference for one style over the others. Determine which is your dominant style, but be open to strategies for all types of learners.

## ▶ Remember–Don't Memorize

You need to use study methods that go beyond rote memorization to genuine comprehension in order to be fully prepared for your test. Using study methods that suit your learning style will help you to *really* learn the material you need to know for the test. One of the most important learning strategies is to be an active reader. Interact with what you are reading by asking questions, making notes, and marking passages instead of simply reading the words on the page. Choose methods of interacting with the text that matches your dominant learning style.

- **Ask questions.** When you read a passage, ask questions such as, "What is the main idea of this section?" Asking yourself questions will test your comprehension of the material. You are also putting the information in your own words, which will help you remember what you have learned. This can be especially helpful when you are learning math techniques. Putting concepts into your own words helps you to understand these processes more clearly.
- **Make notes.** Making notes as you read is another way for you to identify key concepts and to put the material in your own words. Writing down important ideas and mathematical formulas can also help you memorize them.
- **Highlight.** Using a highlighter is another way to interact with what you are reading. Be sure you are not just coloring, but highlighting key concepts that you can return to when you review.
- **Read aloud.** Especially for the auditory learner, reading aloud can help aid in comprehension. Hearing mathematical information and formulas read aloud can clarify their meanings for you.
- **Make connections.** Try to relate what you are reading to things you already know or to a real world example. It might be helpful, for example, to make up a word problem, or draw a diagram or table, to clarify your understanding of what a problem is asking you to do.

Reading actively is probably the most important way to use your study time effectively. If you spend an hour passively reading and retaining little of what you have read, you have wasted that hour. If you take an hour and a half to actively read the same passage, that is time well spent. However, you will not only be learning new material; you will also need methods to review what you have learned.

- **Flashcards.** Just making the cards alone is a way of engaging with the material. You have to identify key concepts, rules, or important information and write them down. Then, when you have made a stack of cards, you have a portable review system. Flashcards are perfect for studying with a friend and for studying on the go.
- **Mnemonics.** These catchy rhymes, songs, and acronyms are tools that help us remember information. Some familiar mnemonics are "i before e except after c" or ROY G. BIV, which stands for Red, Orange, Yellow, Green, Blue, Indigo, Violet—the colors of the rainbow. Developing your own mnemonics will help you make a personal

connection with the material and help you recall it during your test. Mnemonics are also useful when you personalize your "cheat sheet."

- **Personalize your cheat sheet.** Of course, you aren't really going to cheat, but take the Formula Cheat Sheet found on pages xiii–xiv and add to it. Or, highlight the formulas you really need and don't yet know well. This will help them to stand out more than the ones you already know. You can then use the sheet to review—perfect for studying on the go.
- **Outlines and maps.** If you have pages of notes from your active reading, you can create an outline or map of your notes to review. Both tools help you organize and synthesize the material. Most students are familiar with creating outlines using hierarchical headings, but maps may be less familiar. To make a map, write down the main point, idea, or topic under consideration in the middle of a clean piece of paper. Draw a circle around this main topic. Next, draw branches out from that center circle on which to record subtopics and details. Create as many branches as you need—or as many as will fit on your sheet of paper.

## ▶ Studying with Others

Studying in a group or with another person can be a great motivator. It can also be a distraction, as it can be easy to wander off the subject at hand and on to more interesting subjects such as last night's game, or some juicy gossip. The key is to choose your study partners well and to have a plan for the study session that will keep you on track.

There are definite advantages to studying with others:

**Motivation.** If you commit to working with someone else you are more likely to follow through. Also, you may be motivated by some friendly competition.

**Solidarity.** You can draw encouragement from your fellow test takers and you won't feel alone in your efforts. This companionship can help reduce test anxiety.

**Shared expertise.** As you will learn from your practice questions, you have certain strengths and weaknesses in the subject. If you can find a study partner with the opposite strengths and weaknesses, you can each benefit from your partner's strengths. Not only will you get help, but by offering *your* expertise you will build your confidence for the upcoming test.

There are also some disadvantages to studying with others:

**Stress of competition.** Some study partners can be overly competitive, always trying to prove that they are better in the subject than you. This can lead to stress and sap your confidence. Be wary of the overly competitive study partner.

**Too much fun.** If you usually hate studying but really look forward to getting together with your best friend to study, it may be because you spend more time socializing than studying. Sometimes it is better to study with an acquaintance who is well-matched with your study needs and with whom you are more likely to stay on task.

**Time and convenience.** Organizing a study group can take time. If you are spending a lot of time making phone calls and sending e-mails trying to get your study group together, or if you have to travel a distance to meet up with your study partner, this may not be an efficient strategy.

Weigh the pros and cons of studying with others to decide if this is a good strategy for you.

## JUST THE FACTS . . . JUST IN TIME

You have thought about the what, where, when, and how; now you need to put all four factors together to build your study plan. Your study plan should be as detailed and specific as possible. When you have created your study plan, you then need to follow through.

### ▶ Building a Study Plan

You will need a daily planner, a calendar with space to write, or a personal digital assistant to build your plan. You have already determined the time you have free for study. Now you need to fill in the details. You have also figured out what you need to study, and have broken the material down into smaller chunks. Assign one "chunk" of material to each of the longer study sessions you have planned. You may need to combine some "chunks" or add some review sessions depending on the number of long study sessions you have planned in your schedule.

You can also plan how to study in your schedule. For example, you might write for Monday, 6:00 P.M. to 9:00 P.M.: Read chapter four, make notes, map notes, and create set of flashcards. Then for Tuesday, 8:30 A.M. to 9:00 A.M. (your commute time): study chapter four flashcards. The key to a successful study plan is to be as detailed as possible.

## ▶ *Staying on Track*

Bear in mind that nothing goes exactly as planned. You may need to stay late at work, you may get a nasty cold, soccer practice may run late, or your child might need to go to the doctor. Any number of things can happen to your well-thought-out study plan—and some of them probably will. You will need strategies for coping with life's little surprises.

The most important thing to remember when you get off track is not to panic or throw in the towel. You can adjust your schedule to make up the lost time. You may need to reconsider some of your other commitments and see if you can borrow some time for studying. Or you may need to forgo one of your planned review sessions to learn new material. You can always find a few extra minutes here and there for your review.

## ▶ *Minimizing Distractions*

There are some distractions, such as getting sick, that are unavoidable. Many others can be minimized. There are the obvious distractions such as socializing, television, and the telephone. There are also less amusing distractions such as anxiety and fear. They can all eat up your time and throw off your study plan. The good news is you can do a lot to keep these distractions at bay.

- **Enlist the help of your friends and family.** Just as you have asked your friends and family to respect your study space, you can also ask them to respect your study time. Make sure they know how important this test is to you. They will then understand that you don't want to be disturbed during study time, and will do what they can to help you stick to your plan.
- **Keep the television off.** If you know that you have the tendency to get pulled into watching TV, don't turn it on even *before* you plan to study. This way you won't be tempted to push back your study time to see how a program ends or see "what's coming up next."
- **Turn off your cell phone and the ringer on your home phone.** This way you won't eat up your study time answering phone calls— even a five-minute call can cause you to lose focus and waste precious time.
- **Use the relaxation techniques discussed earlier in the chapter** if you find yourself becoming anxious while you study. Breathe, tense and relax your muscles, or visualize a soothing place.
- **Banish negative thoughts.** Negative thoughts—such as, "I'll never get through what I planned to study tonight," "I'm so mad all my friends are at the movies and I'm stuck here studying," "Maybe I'll just study for an hour instead of two so I can watch the season finale

of my favorite show"—interfere with your ability to study effectively. Sometimes just noticing your negative thoughts is enough to conquer them. Simply answer your negative thought with something positive—"If I study the full two hours, I can watch the tape of my show," "I want to study because I want to do well on the test so I can . . . " and so on.

## ▶ Staying Motivated

You can also get off track because your motivation wanes. You may have built a rock-solid study plan and set aside every evening from 6:00 to 9:00 to study. And then your favorite team makes it to the playoffs. Your study plan suddenly clashes with a very compelling distraction. Or you may simply be tired from a long day at work or school or from taking care of your family and feel like you don't have the energy for three hours of concentrated study. Here are some strategies to help keep you motivated:

**Visualization.** Remind yourself of what you will gain from doing well on the test. Take some time to visualize how your life will be positively changed if you accomplish your goal. Do not, however, spend time visualizing how awful your life will be if you fail. Positive visualization is a much more powerful motivator than negative imagery.

**Rewards.** Rewards for staying on track can be a great motivator, especially for flagging enthusiasm. When you accomplish your study goal, perhaps watch your favorite TV program or have a special treat—whatever it is that will motivate you.

**Positive feedback.** You can use your study plan to provide positive feedback. As you work toward the test date, look back at your plan and remind yourself of how much you have already accomplished. Your plan will provide a record of your steady progress as you move forward. You can also enlist the help of study partners, family, and friends to help you stay motivated. Let the people in your life know about your study plan and your progress. They are sure to applaud your efforts.

At the end of the day, *you* will be your prime motivator. The fact that you bought this book and have taken the time to create a well-thought-out study plan shows that you are committed to your goal. As the slogan says, now all that is left is to "Just do it!" Imagine yourself succeeding on your test and let the excitement of meeting your goal carry you forward.

# 2

# Number Properties and Equation Solving

**B**efore you begin learning and reviewing number properties and equation solving, take a few minutes to take this ten-question *Benchmark Quiz*. These questions are similar to the type of questions that you will find on important tests. When you are finished, check the answer key carefully to assess your results. Your Benchmark Quiz analysis will help you determine how much time you need to spend on number properties and equation solving and the specific areas in which you need the most careful review and practice.

## BENCHMARK QUIZ

Please answer the following questions.

1. Evaluate: $-3 + -4(2)$.
   a. $-14$
   b. $14$
   c. $-11$
   d. $-9$
   e. $9$

**2.** Evaluate: $20 - 12 \div (6 - 2)$.
 **a.** 2
 **b.** 16
 **c.** 17
 **d.** 18
 **e.** 20

**3.** Evaluate for $a = -36$, $b = -9$, $c = 5$, and $d = -4$: $a \div b - c - d$
 **a.** 3
 **b.** 3.6
 **c.** 2
 **d.** −5
 **e.** −36

**4.** Solve the equation for $x$: $x - 3 = 12$.
 **a.** 0
 **b.** 4
 **c.** 9
 **d.** 15
 **e.** 36

**5.** Solve the equation for $x$: $\frac{x}{-4} = 11$.
 **a.** −7
 **b.** 7
 **c.** −2.75
 **d.** 44
 **e.** −44

**6.** Solve the equation for $b$: $3b - 11 = 52$.
 **a.** 13.6
 **b.** 17.3
 **c.** 21
 **d.** 28.3
 **e.** 189

**7.** Solve the equation for $c$: $15c - 12 - 3c = 36$.
 **a.** 1
 **b.** 2
 **c.** 2.6
 **d.** 3
 **e.** 4

**8.** Solve the equation for $x$: $8x - 24 = 6x$.
   **a.** −1.7
   **b.** 1.7
   **c.** −12
   **d.** 12
   **e.** 22

**9.** Solve the equation for $p$: $p - 3 = 4(3 - p)$.
   **a.** 2
   **b.** 3
   **c.** 3.75
   **d.** 4
   **e.** 5

**10.** Solve for $a$ in terms of $b$ and $c$: $11a - 6b = c$.

   **a.** $a = \frac{(c + 6b)}{11}$

   **b.** $a = \frac{c}{(6b + 11)}$

   **c.** $a = c + b$

   **d.** $a = \frac{(c + b)}{11}$

   **e.** $a = \frac{(11 + c)}{6b}$

## BENCHMARK QUIZ SOLUTIONS

How did you do on number properties and equation solving? Check your answers here, and then analyze your results to figure out your plan of attack to master these topics.

### ▶ Answers

**1. c.** Order of operations tells you to multiply first; $-3 + -4(2) = -3 + -8$. Then add −3 and −8. The solution is −11.

**2. c.** Evaluate the parentheses first; $20 - 12 ÷ (6 - 2) = 20 - 12 ÷ (4)$. Divide next to get $20 - 3$. Subtract $20 - 3 = 17$ to get the final answer.

**3. a.** First, substitute numbers for the letters: $a = -36$, $b = -9$, $c = 5$, and $d = -4$ into $a ÷ b - c - d$; $-36 ÷ (-9) - 5 - (-4)$. Divide to get $(4) - 5 - (-4)$. Subtract from left to right; $4 - 5 = -1$ so the expression

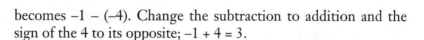

becomes $-1 - (-4)$. Change the subtraction to addition and the sign of the 4 to its opposite; $-1 + 4 = 3$.

**4. d.** Add 3 to both sides of the equation; $x - 3 + 3 = 12 + 3$. Simplify; $x = 15$.

**5. e.** Multiply each side of the equation by $-4$; $-4 \cdot \frac{x}{-4} = 11 \cdot -4$. Since the $-4$'s on the left side cancel out, this leaves $x = -44$.

**6. c.** First add 11 to both sides of the equation; $3b - 11 + 11 = 52 + 11$. This results in $3b = 63$. Divide both sides of the equation by 3; $\frac{3b}{3} = \frac{63}{3}$; $b = 21$.

**7. e.** Combine like terms on the left side of the equation; $12c - 12 = 36$. Add 12 to both sides of the equation; $12c - 12 + 12 = 36 + 12$. This simplifies to $12c = 48$. Divide both sides of the equation by 12; $\frac{12c}{12} = \frac{48}{12}$; $c = 4$.

**8. d.** Subtract $8x$ from both sides of the equation to get the variables on one side; $8x - 8x - 24 = 6x - 8x$. This simplifies to $-24 = -2x$. Divide both sides of the equation by $-2$; $\frac{-24}{-2} = \frac{-2x}{-2}$; $12 = x$.

**9. b.** Use distributive property to get rid of the parentheses; $p - 3 = 12 - 4p$. Add $4p$ to both sides of the equation; $p + 4p - 3 = 12 - 4p + 4p$. Combine like terms; $5p - 3 = 12$. Add 3 to both sides of the equation; $5p - 3 + 3 = 12 + 3$; $5p = 15$. Divide both sides by 5; $\frac{5p}{5} = \frac{15}{5}$; $p = 3$.

**10. a.** Add $6b$ to both sides of the equation; $11a - 6b + 6b = c + 6b$. Simplify; $11a = c + 6b$. Divide both sides of the equation by 11; $a = \frac{(c + 6b)}{11}$.

## BENCHMARK QUIZ RESULTS

If you answered 8–10 questions correctly, you have a good understanding of the properties of numbers, order of operations, and equation solving. After reading through the lesson and focusing on the areas you need to practice on, try the quiz at the end of the chapter to ensure that all of the concepts are clear.

If you answered 4–7 questions correctly, you need to refresh yourself on some of the material. Read through the chapter carefully for review and skill building, and pay careful attention to the sidebars that refer you to more in-depth practice, hints, and shortcuts. Work through the quiz at the end of the chapter to check your progress.

If you answered 1–3 questions correctly, you need help and clarification on the topics in this section. First, carefully read this chapter and concentrate on these basic number skills. Perhaps you learned this information once and forgot it, so take the time now to refresh your skills and improve your knowledge. After taking the quiz at the end of the chapter, you may want to reference a more in-depth and comprehensive book, such as Lessons 1 and 2 in LearningExpress's *Algebra Success in 20 Minutes a Day* or practice more problems in *501 Algebra Questions*.

## JUST IN TIME LESSON—NUMBER PROPERTIES AND EQUATION SOLVING

This lesson covers the basics of working with algebraic expressions and equations.

Topics include:

- the properties of numbers that help you to simplify and solve using algebra
- operations with integers (positive and negative numbers)
- the order of operations of numerical expressions
- equation solving.

## ▶ Properties of Numbers

Although the actual names of the properties may not be tested, you need to be familiar with the ways each one helps to simplify problems and solve equations. You will also notice that most properties work for addition and multiplication, but not for subtraction and division.

**Commutative Property**: This property states that even though the order of the numbers changes, the answer is the same. This property works under addition and multiplication.

*Examples:*

$a + b = b + a$      $ab = ba$

$3 + 4 = 4 + 3$      $3 \bullet 4 = 4 \bullet 3$

SHORTCUT

Commutative sounds like the word *commute*. Think of a commute from home to school or work and then from school or work back to home. Even though the order is different, the distance is still the same. If you change the order of the numbers using the commutative property, the answer is still the same.

**Associative Property**: This property states that even though the grouping of the numbers changes, the result or answer is the same. This property also works under addition and multiplication.

*Examples:*

$a + (b + c) = (a + b) + c$     $a(bc) = (ab)c$

$2 + (3 + 5) = (2 + 3) + 5$     $2 \cdot (3 \cdot 5) = (2 \cdot 3) \cdot 5$

SHORTCUT

When you *associate* with different people, you are grouped with them. Remember that associative property changes the grouping of numbers within parentheses.

**Identity Property**

*Addition:* Any number plus zero is itself. Zero is the additive identity element.

$a + 0 = a$       $5 + 0 = 5$

*Multiplication:* Any number times one is itself. One is the multiplicative identity element.

$a \cdot 1 = a$       $5 \cdot 1 = 5$

**Inverse Property**: This property is often used in equation solving when you want a number to cancel out.

*Addition:* The additive inverse of any number is its opposite.

$a + (-a) = 0$       $3 + (-3) = 0$

*Multiplication:* The multiplicative inverse of any number is its reciprocal.

$a \cdot \frac{1}{a} = 1$       $6 \cdot \frac{1}{6} = 1$

**Distributive Property:** This property is used when there are two different operations, multiplication and addition or multiplication and subtraction. Basically it states that the number being multiplied must be multiplied, or distributed, to each term within the parentheses. You will use this property when solving most equations with parentheses.

$a (b + c) = ab + ac$     or     $a (b - c) = ab - ac$
$5(a + 2) = 5 \bullet a + 5 \bullet 2$ which simplifies to $5a + 10$
$2(3x - 4) = 2 \bullet 3x - 2 \bullet 4$ which simplifies to $6x - 8$

## ▶ Integers and Absolute Value

The absolute value of a number is the distance a number is away from zero on a number line. The symbol for absolute value is two bars surrounding the number or expression. Absolute value is always positive because it is a measure of distance.

$|5| = 5$ because 5 is five units from zero on a number line.
$|-3| = 3$ because $-3$ is three units from zero on a number line.

## GLOSSARY
**INTEGERS** the set of whole numbers and their opposites with zero being its own opposite; $\{ \dots , -2, -1, 0, 1, 2, 3, \dots \}$.

Integers are really just positive and negative whole numbers. When performing any operations with integers, use the absolute value of the number and then determine the sign of your answer. Here are some helpful rules to follow.

### Adding and Subtracting Integers
1. If adding and the signs are the same, add the absolute value of the numbers and keep the sign.

    **a.** $3 + 4 = 7$      **b.** $-2 + -13 = -15$

2. If adding and the signs are different, subtract the absolute value of the numbers and take the sign of the number with the larger absolute value.

    **a.** $-5 + 8 = 3$      **b.** $10 + -14 = -4$

3. If subtracting, change the subtraction sign to addition and change the sign of the number following to its opposite. Then follow the rules for addition

**a.** $-5 - 6$       **b.** $-12 - (-7)$
    $-5 + -6$          $-12 + (+7)$
    $-11$             $-5$

 SHORTCUT

**If there is no sign in front of a number or variable, it is positive.**

## Multiplying and Dividing Integers

1. If there are an even number of negatives, multiply or divide as usual and the answer is positive. Remember that zero is an even number so if there are no negatives, the answer is positive.

   **a.** $-3 \bullet -4 = 12$    **b.** $(-12 \div -6) \bullet 3 = 6$

2. If there are an odd number of negatives, multiply or divide as usual and the answer is negative.

   **a.** $-15 \div 5 = -3$    **b.** $(-2 \bullet -4) \bullet -5 = -40$

 EXTRA HELP

Try the website www.aaamath.com for more information and practice using integers. At this site you will find lessons and explanations, along with games to increase your math skills. For even more help, see *Algebra Success in 20 Minutes a Day*, Lesson 1: Working with Integers.

## ▶ *Order of Operations*

There is a specific order in which to complete the operations in a multi-step expression. This particular order can be remembered as **P**lease **E**xcuse **M**y **D**ear **A**unt **S**ally, or **PEMDAS**. In any expression, evaluate in this order:

**P**    **P**arentheses/grouping symbols first
**E**    then **E**xponents
**MD** **M**ultiplication/**D**ivision in order from left to right
**AS**   **A**ddition/**S**ubtraction in order from left to right

Keep in mind that division may be done before multiplication, and subtraction may be done before addition, depending on which operation is first when working from left to right. There are two exceptions:

1. A fraction bar acts as a grouping symbol. In the expression $\frac{3+7}{2}$, addition is done first.

2. In a problem such as $4y \div 2x$, it is implied that $4y$ and $2x$ be simplified (multiplied) first before division.

*Examples:*

Evaluate the following using order of operations:

**a.** $2 \bullet 3 + 4 - 2$

| | |
|---|---|
| $6 + 4 - 2$ | Multiply first. |
| $10 - 2$ | Add and subtract in order from left to right. |
| $8$ | |

**b.** $3^2 - 16 + (5 - 1)$

| | |
|---|---|
| $3^2 - 16 + (4)$ | Evaluate parentheses first. |
| $9 - 16 + 4$ | Evaluate exponents. |
| $-7 + 4$ | Subtract and then add in order from left to right. |
| $-3$ | |

**c.** $[2(4^2 - 9) + 3] - 1$

| | |
|---|---|
| $[2(16 - 9) + 3] - 1$ | Begin with the innermost grouping sym |
| $[2(7) + 3] - 1$ | bols and follow **PEMDAS** (Here, exponents are first within the parentheses). Continue with order of operations, working from the inside out (subtract within the parentheses). |
| $[14 + 3] - 1$ | Multiply. |
| $[17] - 1$ | Add $14 + 3$. |
| $16$ | Subtract to complete the problem. |

**d.** Evaluate $ab + c$ for $a = 2$, $b = -3$, and $c = 12$.

In a question such as this, first substitute the numbers in for each letter of the problem; $ab + c$ then becomes $2 \bullet -3 + 12$

 RULE BOOK

**If there is not an operation written between two variables, they should be multiplied (ab means "a times b").**

Now, follow the order of operations.

| | |
|---|---|
| $-6 + 12$ | Multiply. |
| $6$ | Add to complete the problem. |

Two types of problems that often appear on standardized tests that also apply the principle of order of operations are functions and special types of defined operations.

## ▶ Functions

Functions are a special type of equation often in the form $f(x)$. Suppose you are given a function such as $f(x) = 3x + 2$. To evaluate $f(4)$, substitute 4 into the function for $x$ and use the correct order of operations.

$$f(x) = 3x + 2$$
$$f(4) = 3(4) + 2$$
$$= 12 + 2$$
$$= 14$$

 EXTRA HELP

For more information and practice on order of operations, see *Algebra Success in 20 Minutes a Day*, Lesson 2: Working with Algebraic Expressions.

## ▶ Special Types of Defined Operations

There may be some unfamiliar operations that appear on your standardized test. These questions may involve operations that use symbols like #, $, &, or @. Usually these problems are solved by simple substitution, and will only really involve operations that you do know.

*Example:*
For $a \# b$ defined as $a^2 - 2b$, what is the value of $3 \# 2$?

For this question, use the definition of the operation as the formula and substitute in the values of 3 and 2 for $a$ and $b$, respectively; $a^2 - 2b = 3^2 - 2(2) = 9 - 4 = 5$.

## ▶ Solving Linear Equations of One Variable

The goal when solving any equation is to get the variable alone.

 RULE BOOK

To get the variable alone perform the inverse, or opposite, operation of the number you want to eliminate on both sides of the equation.

When solving this type of equation, it is important to remember two basic properties:

• If a number is added to or subtracted from one side of an equation, it must be added or subtracted to the other side.

- If a number is multiplied or divided on one side of an equation, it must also be multiplied or divided on the other side.

 RULE BOOK

**Whatever you do to one side of the equal sign needs to also be done on the other side.**

Equation solving with linear equations has four basic steps:

1. Remove parentheses by using distributive property.
2. Combine like terms on the same side of the equal sign.
3. Move the variables to one side of the equation.
4. Solve the one- or two-step equation that remains, remembering the two properties mentioned above.

 RULE BOOK

**The goal of solving any equation is to isolate the variable. In other words, get the variable *alone*.**

*Examples:*
Solve each of the following for $x$:

**a.** This is an example of a one-step equation.
Solve for $x$: $\frac{x}{3} = 9$.
Multiply both sides of the equation by 3 to get $x$ alone;
$3 \bullet \frac{x}{3} = 9 \bullet 3$
$x = 27$

 SHORTCUT

**Remember that $\frac{x}{4} = \frac{1}{4}x$ and $\frac{2x}{3} = \frac{2}{3}x$.**

**b.** This is an example of a two-step equation.
Solve for $x$: $3x - 5 = 10$.
Add five to both sides of the equation: $3x - 5 + 5 = 10 + 5$
Divide both sides by 3: $\frac{3x}{3} = \frac{15}{3}$
$x = 5$

## CALCULATOR TIP

Most graphing calculators have a **Solver** function that will solve equations for you. This function is often under that **Math** menu. Consult your calculator manual on how to use this function on your particular calculator. Remember that if there is more than one solution, the calculator only finds them one at a time.

**a.** This is an example containing parentheses.
  Solve for $x$: $3(x - 1) + x = 1$.
  Use distributive property to remove parentheses: $3x - 3 + x = 1$
  Combine like terms: $4x - 3 = 1$
  Add 3 to both sides of the equation: $4x - 3 + 3 = 1 + 3$
  Divide both sides by 4: $\frac{4x}{4} = \frac{4}{4}$
  $$x = 1$$

**b.** This is an example of variables on both sides of the equal sign.
  Solve for $x$: $8x - 2 = 8 + 3x$.
  Subtract $3x$ from both sides of the equation to move the variables
    to one side: $8x - 3x - 2 = 8 + 3x - 3x$
  Add 2 to both sides of the equation: $5x - 2 + 2 = 8 + 2$
  Divide both sides by 5: $\frac{5x}{5} = \frac{10}{5}$
  $x = 2$

## EXTRA HELP

For more information and practice on solving equations, see *Algebra Success in 20 Minutes a Day*, Lesson 4: Solving Basic Equations, Lesson 5: Solving Multi-step Equations, and Lesson 6: Solving Equations with Variables on Both Sides of the Equation.

## ▶ *Solving Literal Equations and Formulas*

Sometimes equation solving will take the form of a problem containing *many* variables and numbers. A literal equation is an equation that contains two or more variables. It may be in the form of a formula. You may be asked to solve a literal equation for one variable in terms of the other variables. The steps to do this are the same as solving linear equations.

*Example:*
Solve for $x$ in terms of $a$ and $b$: $2x + b = a$
Subtract $b$ from both sides of the equation: $2x + b - b = a - b$
Divide both sides of the equation by 2: $\frac{2x}{2} = \frac{a-b}{2}$
$x = \frac{a-b}{2}$

Many times questions will take the form of real-life formulas, such as distance, interest, and temperature. To solve this type of question, simply substitute any of the known values into the given formula and solve as you would a regular equation.

*Example:*
Using the formula *distance = rate × time*, find the speed of a car that travels 110 miles in $2\frac{1}{2}$ hours.
Use $d = r \times t$ as the formula. Note that distance in this case is the miles traveled and the rate is the speed of the car. Substitute $d = 110$ and $t = 2.5$ and solve: $d = r \times t$; $110 = r \times 2.5$.
Divide both sides of the equation by 2.5; $\frac{110}{2.5} = \frac{2.5r}{2.5}$; $44 = r$.
The speed of the car is 44 miles per hour.

## EXTRA HELP

**For more information and practice on solving formula questions, see** *Algebra Success in 20 Minutes a Day*, **Lesson 7: Using Formulas to Solve Equations.**

## TIPS AND STRATEGIES

Evaluating expressions and solving equations on most standardized tests are not any more difficult than what you have encountered in middle school and high school math. Here are a couple of things to keep in mind.

- Apply the rules for operations with integers in all questions containing positive and negative numbers.
- Remember to always use the correct order of operations when evaluating a mathematical expression.
- Many questions with variables, such as formulas, can be solved rather easily by substitution. If stuck on a particular question, see if you can substitute numbers in order to simplify.
- Use the properties of numbers to help you simplify and solve. For example, use the inverse property of multiplication to eliminate any fractions with the variable.

$$\tfrac{1}{4}x = 6$$
$$4 \bullet \tfrac{1}{4}x = 6 \bullet 4$$
$$x = 24$$

- When dealing with equations, be sure to use the inverse operation of what you are trying to eliminate.

- In any equation, whatever is done to one side of the equal sign must be done to the other side.

- Don't forget that the goal for solving an equation is to get the variable alone. This is also called "isolating the variable."

- Keep in mind that $\tfrac{x}{2}$ can also be written as $\tfrac{1}{2}x$. This may be a time saver and make some problems appear less complicated.

- Subtracting a number is the same as adding its opposite. Problems may seem easier if you change subtraction to addition and change the sign of the number that follows to its opposite.

- Remember that solving problems that involve formulas can be as simple as substituting numbers and simplifying. Many times in this type of question the formula will be given to you.

- Be on the lookout for "traps" in multiple-choice questions. The incorrect answer choices will be the result of making common errors. Work out most problems and check your answer before looking at the answer choices.

## CHAPTER QUIZ

Try these practice problems to track your progress through properties and equation solving.

1. Evaluate: $5 + 6 \bullet 2$.
   a. 9
   b. 11
   c. 13
   d. 17
   e. 22

2. Evaluate: $12 - 3 + 4$.
   a. 5
   b. 8
   c. 13
   d. 0
   e. 19

**3.** Evaluate: $21 - (6 - 2) \div 2$.
   **a.** 6.5
   **b.** 8.5
   **c.** 15
   **d.** 19
   **e.** 34

**4.** Evaluate for $x = 10$ and $y = -5$: $y + x \div y$.
   **a.** $-1$
   **b.** 1
   **c.** $-3$
   **d.** 3
   **e.** $-7$

**5.** Evaluate for $a = -3$, $b = 5$ and $c = -1$: $ab - bc + 1$.
   **a.** $-19$
   **b.** $-11$
   **c.** $-9$
   **d.** 11
   **e.** 21

**6.** Solve the equation for $x$: $x + 5 = 10$.
   **a.** 5
   **b.** 14
   **c.** 40
   **d.** 2.5
   **e.** 13

**7.** Solve the equation for $y$: $-6y = 30$.
   **a.** 35
   **b.** 25
   **c.** $-5$
   **d.** $-35$
   **e.** 5

**8.** Solve the equation for $t$: $\frac{t}{-10} = -90$.
   **a.** $-9$
   **b.** 9
   **c.** $-80$
   **d.** $-900$
   **e.** 900

9. Solve the equation for $y$: $y - 16 = -72$.
   a. $-88$
   b. $-78$
   c. $-56$
   d. $-4.5$
   e. $4.5$

10. Solve the equation for $x$: $2x + 3 = 15$.
    a. $10$
    b. $8$
    c. $6$
    d. $4$
    e. $2$

11. Solve the equation for $x$: $-x - 9 = 6$.
    a. $15$
    b. $-15$
    c. $-5$
    d. $5$
    e. $-10$

12. Solve the equation for $h$: $\frac{h}{3} + 9 = -18$.
    a. $-81$
    b. $-15$
    c. $-9$
    d. $9$
    e. $81$

13. Solve the equation for $a$: $6a + 9a = 90$.
    a. $2.5$
    b. $3$
    c. $6$
    d. $10$
    e. $30$

14. Solve the equation for $x$: $-20x - 8x + 1 = 57$.
    a. $-2$
    b. $-7$
    c. $2$
    d. $7$
    e. $-3.2$

**15.** Solve the equation for $x$: $3x - 8 = 5x$.
  **a.** 1
  **b.** 2
  **c.** –2
  **d.** –4
  **e.** 4

**16.** Solve the equation for $a$: $9a + 12 = 6a - 12$.
  **a.** –8
  **b.** –6
  **c.** –2.4
  **d.** 6
  **e.** 8

**17.** Solve the equation for $p$: $9p + 12 - 6p + 2 = 14$.
  **a.** 4
  **b.** $1\frac{1}{15}$
  **c.** 3
  **d.** 0
  **e.** no solution

**18.** Solve the equation for $x$: $\frac{1}{4}(x + 4) = 10$.
  **a.** 5
  **b.** 9
  **c.** 24
  **d.** 36
  **e.** 40

**19.** Solve the equation for $x$: $0.2(x - 3) + 0.3 = 4.5$.
  **a.** 1.8
  **b.** 2.4
  **c.** 18
  **d.** 24
  **e.** 30

**20.** Solve the equation for $x$: $\frac{3(x + 4)}{-4} = 6$.
  **a.** –12
  **b.** 2.3
  **c.** –18
  **d.** –2
  **e.** 2

**21.** Solve the equation for $x$: $-4(x + 8) + 7x = 2x + 32$.
   **a.** 0
   **b.** 6.4
   **c.** 16
   **d.** 32
   **e.** 64

**22.** Solve the equation for $w$ in terms of $A$ and $l$: $A = lw$.
   **a.** $w = Al$
   **b.** $w = A - l$
   **c.** $w = A + l$
   **d.** $w = \frac{A}{l}$
   **e.** $w = 2Al$

**23.** Solve the equation for $a$ in terms of $b$ and $c$: $7ab = c$.
   **a.** $a = 7bc$
   **b.** $a = \frac{bc}{7}$
   **c.** $a = \frac{c}{7b}$
   **d.** $a = \frac{7c}{b}$
   **e.** $a = \frac{b}{7c}$

**24.** Solve the equation for $j$ in terms of $h$ and $k$: $3j + h = k$.
   **a.** $j = \frac{k + h}{3}$
   **b.** $j = \frac{k - h}{3}$
   **c.** $j = 3k - h$
   **d.** $j = k - 3h$
   **e.** $j = \frac{3}{k - h}$

**25.** Using the formula *Interest = principle × rate × time* ($I = prt$), what is the interest earned on a savings account with a balance of $1,500 when the interest rate is 5% for 4 years?
   **a.** $50
   **b.** $300
   **c.** $375
   **d.** $1,800
   **e.** $30,000

## ANSWERS

Here are the answers and explanations for the chapter quiz. Read over the explanations carefully to correct any misunderstandings. Refer to Learning-Express's *Algebra Success in 20 Minutes A Day* for further review and practice.

**1. d.** Perform multiplication first; $5 + 6 \cdot 2 = 5 + 12$. Add to complete the problem. The solution is 17.

**2. c.** Since there are no other operations, add and subtract in order from left to right; $12 - 3 + 4 = 9 + 4 = 13$.

**3. d.** Evaluate the parentheses first; $21 - (6 - 2) \div 2 = 21 - (4) \div 2$. Divide to get $21 - 2$. Subtract to get a final answer of 19.

**4. e.** Substitute the values of $x = 10$ and $y = -5$ into the expression $y + x \div y$. This becomes $-5 + 10 \div -5$. Divide 10 and $-5$ to get $-5 + -2$. Add. Note that the signs are the same so add the absolute values and keep the negative sign; $-5 + -2 = -7$.

**5. c.** Substitute the values of $a = -3$, $b = 5$ and $c = -1$ into the expression $ab - bc + 1$. This becomes $(-3)(5) - (5)(-1) + 1$. Perform multiplication to get $(-15) - (-5) + 1$. Change subtraction to addition and the sign of $-5$ to $+5$; $-15 + 5 + 1$. Add in order from left to right; $-10 + 1 = -9$.

**6. a.** Subtract 5 from both sides of the equation; $x + 5 - 5 = 10 - 5$. The variable is now alone; $x = 5$.

**7. c.** Divide each side of the equation by $-6$; $\frac{-6y}{-6} = \frac{30}{-6}$. The variable is now alone; $y = -5$.

**8. e.** Multiply each side by $-10$; $-10 \cdot \frac{t}{-10} = -90 \cdot -10$. The $-10$'s on the left side cancel out and the equation becomes $t = -90 \cdot -10 = 900$; $t = 900$.

**9. c.** Add 16 to both sides of the equation; $y - 16 + 16 = -72 + 16$. This simplifies to $y = -72 + 16$. So, $y = -56$.

**10. c.** Subtract 3 from both sides of the equation; $2x + 3 - 3 = 15 - 3$. Divide each side of the equation by 2; $\frac{2x}{2} = \frac{12}{2}$. The variable is now alone; $x = 6$.

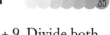

**11. b.** Add 9 to both sides of the equation; $-x - 9 + 9 = 6 + 9$. Divide both sides of the equation by $-1$. Remember that $-x = -1x$; $\frac{-x}{-1} = \frac{15}{-1}$. The variable is now alone; $x = -15$.

**12. a.** Subtract 9 from both sides of the equation; $\frac{b}{3} + 9 - 9 = -18 - 9$. The equation becomes $\frac{b}{3} = -27$. Multiply both sides of the equation by 3; $3 \bullet \frac{b}{3} = -27 \bullet 3$; $b = -81$.

**13. c.** Combine like terms on the left side of the equation; $15a = 90$. Divide each side of the equation by 15; $\frac{15a}{15} = \frac{90}{15}$. The variable is now alone; $a = 6$.

**14. a.** Change subtraction to addition and the sign of the following term to its opposite; $-20x + -8x + 1 = 57$. Combine like terms on the left side of the equation; $-28x + 1 = 57$. Subtract 1 from both sides of the equation; $-28x + 1 - 1 = 57 - 1$. Divide each side of the equation by $-28$; $\frac{-28x}{-28} = \frac{56}{-28}$. The variable is now alone; $x = -2$.

**15. d.** Subtract $3x$ from both sides to get the variable on one side of the equation; $3x - 3x - 8 = 5x - 3x$. The equation simplifies to $-8 = 2x$. Divide both sides by 2 to get $x$ alone; $\frac{-8}{2} = \frac{2x}{2}$; $-4 = x$.

**16. a.** Subtract $6a$ from both sides of the equation to get the variable on one side; $9a - 6a + 12 = 6a - 6a - 12$. The equation simplifies to $3a + 12 = -12$. Subtract 12 from both sides; $3a + 12 - 12 = -12 - 12$. So, $3a = -24$. Divide both sides of the equation by 3; $\frac{3a}{3} = \frac{-24}{3}$; $a = -8$.

**17. d.** Use commutative property to arrange like terms; $9p - 6p + 12 + 2 = 14$. Combine like terms on the left side of the equation; $3p + 14 = 14$. Subtract 14 from both sides of the equation; $3p + 14 - 14 = 14 - 14$. Divide both sides by 3; $\frac{3p}{3} = \frac{0}{3}$; $p = 0$.

**18. d.** Use distributive property on the left side of the equation; $\frac{1}{4} \bullet x + \frac{1}{4} \bullet 4 = 10$. Remember that $\frac{1}{4} \bullet 4 = 1$. Subtract 1 from both sides of the equation; $\frac{1}{4}x + 1 - 1 = 10 - 1$. Multiplying by the reciprocal of a fraction is the same as dividing by that fraction. Multiply each side of the equation by $\frac{4}{1}$; $\frac{4}{1} \bullet \frac{1}{4}x = \frac{4}{1} \bullet 9$. The variable is now alone; $x = 36$.

Another approach to this equation would be to eliminate the $\frac{1}{4}$ first by multiplying each side of the equation by 4; $4 \bullet \frac{1}{4}(x + 4) = 4$

• 10. The next step is to subtract 4 on both sides of the equation; $x + 4 - 4 = 40 - 4$. The variable is now alone; $x = 36$.

**19. d.** Use distributive property on the left side of the equation; $0.2x - 0.6 + 0.3 = 4.5$. Change subtraction to addition and the sign of the following number to its opposite; $0.2x + -0.6 + 0.3 = 4.5$. Combine like terms on the left side of the equation; $0.2x + -0.3 = 4.5$. Add 0.3 to both sides of the equation; $0.2x + -0.3 + 0.3 = 4.5 + 0.3$. This simplifies to $0.2x = 4.8$. Divide both sides by 0.2; $x = 24$.

**20. a.** There are a few different ways to approach solving this problem.

Method I:
Multiply both sides of the equation by $-4$; $-4 \cdot \frac{3(x + 4)}{-4} = 6 \cdot -4$. This simplifies to $3(x + 4) = -24$. Divide each side of the equation by 3; $\frac{3(x + 4)}{3} = \frac{-24}{3}$. This simplifies to $x + 4 = -8$. Subtract 4 from both sides of the equation; $x + 4 - 4 = -8 - 4$. The variable is now alone; $x = -12$.

Method II:
Another way to look at the problem is to still multiply each side by $-4$ in the first step to get $3(x + 4) = -24$. Then use distributive property on the left side; $3x + 12 = -24$. Subtract 12 from both sides of the equation; $3x + 12 - 12 = -24 - 12$. Simplify. $3x = -36$. Divide each side by 3; $\frac{3x}{3} = \frac{-36}{3}$. The variable is now alone; $x = -12$.

**21. e.** Use distributive property on both sides of the equation; $-4x - 32 + 7x = 2x + 32$. Use commutative property to arrange like terms; $-4x + 7x - 32 = 2x + 32$. Combine like terms on the left side of the equation; $3x - 32 = 2x + 32$. Subtract $2x$ from both sides of the equation; $3x - 2x - 32 = 2x - 2x + 32$. Simplify. (Remember $1x = x$.) $x - 32 = 32$. Add 32 to both sides of the equation; $x - 32 + 32 = 32 + 32$; $x = 64$.

**22. d.** To solve for $w$, divide both sides by $l$; $\frac{A}{l} = \frac{lw}{l}$. The variable $w$ is now alone; $\frac{A}{l} = w$. Remember, this is the same result as $w = \frac{A}{l}$.

**23. c.** To get the variable $a$ alone, divide both sides of the equation by $7b$; $\frac{7ab}{7b} = \frac{c}{7b}$. Both $7b$'s on the left side cancel, leaving only $a$; $a = \frac{c}{7b}$.

**24. b.** To get $j$ alone, first subtract $h$ from both sides of the equation. This results in $3j + h - h = k - h$. This equation simplifies to $3j = k - h$. Divide both sides by 3; $\frac{3j}{3} = \frac{k-h}{3}$. So, $j = \frac{k-h}{3}$.

**25. b.** Use the formula $I = prt$ and substitute $p = 1{,}500$, $r = 0.05$ (the percent written as a decimal) and $t = 4$; $I = (1{,}500)(0.05)(4)$. By multiplying, the result is 300. This account made $300 in interest over the 4 years.

# 3

# Coordinate Geometry and Graphing Linear Equations

**B**efore you begin learning and reviewing coordinate geometry and graphing applications, take a few minutes to take this ten-question *Benchmark Quiz*. These questions are similar to the type of questions that you will find on important tests. When you are finished, check the answer key carefully to assess your results. Your Benchmark Quiz analysis will help you determine how much time you need to spend on this chapter and the specific areas in which you need the most careful review and practice.

## BENCHMARK QUIZ

1. In what quadrant is the point (–1,2) located?
   a. I
   b. II
   c. III
   d. IV
   e. None of these

**2.** Point $S$ in the graph below is located at

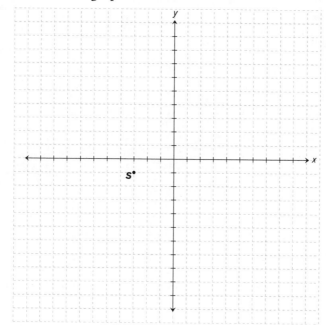

   **a.** $(1,3)$
   **b.** $(-1,3)$
   **c.** $(3,-1)$
   **d.** $(-3,-1)$
   **e.** $(3,1)$

**3.** What is the slope-intercept form of the equation $2x - 3y = 9$?
   **a.** $y = -\frac{2}{3}x - 3$

   **b.** $y = \frac{2}{3}x + 3$

   **c.** $y = \frac{2}{3}x - 3$

   **d.** $y = \frac{3}{2}x + 2$

   **e.** $y = -\frac{3}{2}x - 2$

**4.** What is the slope of the line $y = -\frac{1}{3}x - 4$?
   **a.** $-4$
   **b.** $-3$
   **c.** $-\frac{1}{4}$
   **d.** $-\frac{1}{3}$
   **e.** $-1$

**5.** What is the y-intercept of the line $2y = x - 2$?
  **a.** $-2$
  **b.** $-\frac{1}{2}$
  **c.** $-1$
  **d.** $\frac{1}{2}$
  **e.** $2$

**6.** What is the slope of the line $x = -3$?
  **a.** $0$
  **b.** $3$
  **c.** $-3$
  **d.** $1$
  **e.** undefined

**7.** Which of the following points is a solution to the graph $2x + 3y = 8$?
  **a.** $(1, 8)$
  **b.** $(2, 2)$
  **c.** $(1, 3)$
  **d.** $(1, 2)$
  **e.** $(2, 1)$

**8.** What is the equation of the graph in the figure below?

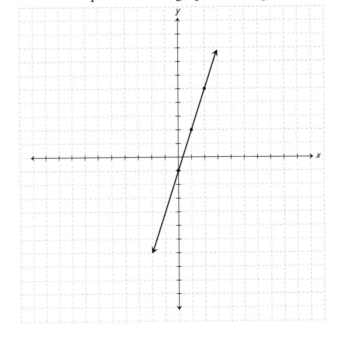

a. $y = 3x - 1$
b. $y = -3x + 1$
c. $y = \frac{1}{3}x + 1$
d. $y = -\frac{1}{3}x + 1$
e. $y = \frac{1}{3}x - 1$

9. Which of the following graphs represents the line $y = -2x + 2$?

a.

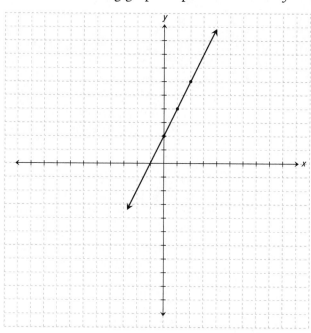

b.

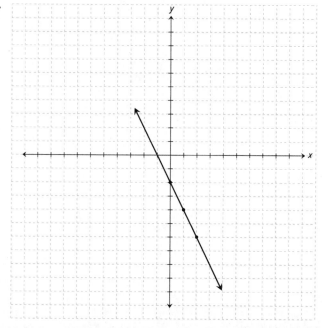

**c.**

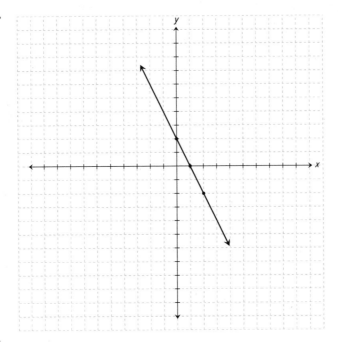

**d.**

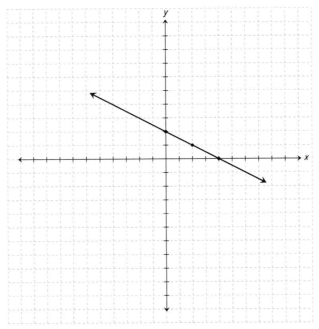

e.

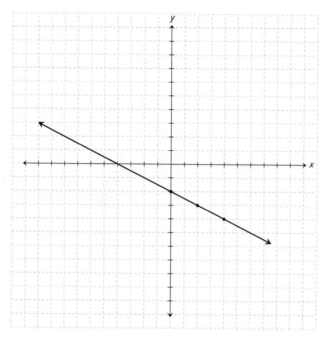

**10.** What is the slope of the line that contains the points (–2,3) and (3,4)?

a. –5
b. $-\frac{1}{5}$
c. 0
d. $\frac{1}{5}$
e. 5

## BENCHMARK QUIZ SOLUTIONS

How did you do on coordinate geometry and graphing linear equations? Check your answers here, and then analyze your results to figure out your plan of attack to master these topics.

## ▶ *Answers*

**1. b.** To find the location of the point (–1,2), start at the origin and move 1 unit to the left. From that point, move 2 units up. Moving over to the left and up puts the point in quadrant II.

**2. d.** To get to point $S$, start at the origin and move 3 units to the left. Moving to the left is a negative direction, so the $x$-coordinate is –3. From there, move 1 unit down. Moving down is also a negative direction, so the $y$ coordinate is –1. Point $S$ is located at $(-3,-1)$.

**3. c.** Slope-intercept form of an equation is $y = mx + b$. To change the equation $2x - 3y = 9$, first subtract $2x$ from both sides; $2x - 2x - 3y = 9 - 2x$. This simplifies to $-3y = 9 - 2x$. Use commutative property to switch $-2x$ and 9 since you are looking for $mx + b$ form. Divide both sides by –3; $\frac{-3y}{-3} = \frac{-2x}{-3} + \frac{9}{-3}$. This simplifies to $y = \frac{2}{3}x - 3$.

**4. d.** This equation is in slope-intercept form, which is written as $y = mx + b$. The number $m$, or coefficient, in front of $x$ represents the slope of the line. The slope is $-\frac{1}{3}$.

**5. c.** To find the $y$-intercept of the equation, first change to $y = mx + b$ form. To get $y$ alone, divide both sides of the equation by 2; $\frac{2y}{2} = \frac{x}{2} - \frac{2}{2}$. This simplifies to $y = \frac{1}{2}x - 1$ (Recall that $\frac{x}{2} = \frac{1}{2}x$). The $y$-intercept is $b$ from the equation, which is –1.

**6. e.** The equation $x = -3$ is a vertical line. The slope of any vertical line is undefined.

**7. d.** Substitute the $x$-values from the answer choices to see what the corresponding $y$-value is. Since three of the answer choices have an $x$-value of 1, try that first. Substitute 1 in for $x$; $2x + 3y = 8$ becomes $2(1) + 3y = 8$. Simplify to get $2 + 3y = 8$. Subtract 2 from both sides of the equation; $2 - 2 + 3y = 8 - 2$. This simplifies to $3y = 6$. Divide both sides by 3; $\frac{3y}{3} = \frac{6}{3}$. So, $y = 2$. Therefore, a point on this line is $(1,2)$.

**8. a.** To find the equation of the line, find the slope and the $y$-intercept of the line and write it in $y = mx + b$ form. Remember that $m$ equals the slope of the line and $b$ equals the $y$-intercept of the line. Since the line crosses the $y$-axis at the point $(0,-1)$, the $y$-intercept, or $b$, is –1. To move from one point to another on the graph, you need to move up 3 units and over 1 unit. This is a slope of $\frac{3}{1}$, which is equal to 3. An equation with a slope of 3 and a $y$-intercept of –1 is written as $y = 3x - 1$.

**9. c.** The equation $y = -2x + 2$ has a slope of $-2$ and a $y$-intercept of 2. Answer choice **c** crosses the $y$-axis at 2, which is the $y$-intercept. To move from one point to another on that line, go down 2 units and over 1 to the right. This is a slope of $-2$. Choice **c** represents the equation $y = -2x + 2$.

**10. d.** Use the slope formula to find the difference in the $y$ values over the difference in the $x$ values. The slope is equal to $\frac{\text{change in } y}{\text{change in } x} = \frac{y_1 - y_2}{x_1 - x_2} = \frac{3 - 4}{-2 - 3} = \frac{-1}{-5} = \frac{1}{5}.$

## BENCHMARK QUIZ RESULTS

If you answered 8–10 questions correctly, you have a good understanding of coordinate geometry and graphing linear functions. After reading through the lesson and focusing on the areas you need to review, try the quiz at the end of the chapter to ensure that all of the concepts are clear.

If you answered 4–7 questions correctly, you need to refresh yourself on some of the material. Read through the chapter carefully for review and skill building, and pay careful attention to the sidebars that refer you to more in-depth practice, hints, and shortcuts. Work through the quiz at the end of the chapter to check your progress.

If you answered 1–3 questions correctly, you need help and clarification on the topics in this section. First, carefully read this chapter and concentrate on these coordinate geometry and linear equation concepts. Perhaps you learned this information and forgot, so take the time now to refresh your skills and improve your knowledge. After taking the quiz at the end of the chapter, you may want to reference a more in-depth and comprehensive book, such as LearningExpress's *Algebra Success in 20 Minutes a Day*.

## JUST IN TIME LESSON—COORDINATE GEOMETRY AND GRAPHING LINEAR EQUATIONS

This lesson covers the key components of coordinate geometry and graphing in the coordinate plane. You will review how to:

- plot points
- use slope, midpoint, and distance formulas
- graph equations by table method and slope-intercept method

## ▶ Coordinate Graphing

 GLOSSARY

**COORDINATE PLANE** the region created by the intersection of two perpendicular signed number lines, the *x*- and *y*-axes. The plane is divided into four quadrants that are numbered I, II, III, and IV as shown in the graph below.

**ORIGIN** the location in the coordinate plane where the *x*-axis and *y*-axis intersect. It is located at point **A** in the graph below. This point (0,0) can be used as the starting point when graphing coordinates.

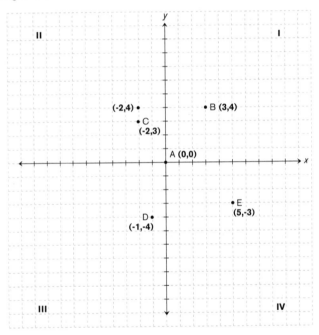

Each location in the plane is named by a point (*x*, *y*) and these numbers are called the coordinates of the point. For example, the point (–2,4) has an *x*-coordinate of –2 and a *y*-coordinate of 4. Each point is found by starting at the intersection of the axes, or the origin, and moving *x* units to the right or left and then *y* units up or down. Positive directions are to the right and up and negative directions are to the left and down.

## Examples of Graphing Points

Here are some examples on how to graph points located in different quadrants.

1. To graph the point (3,4), start at the origin. Go to the right 3 units and from there go up 4 units. This is point **B** in the graph on the previous page, located in quadrant I.
2. To graph the point (–2,3), start at the origin. Go to the left 2 units and from there go up 3 units. This is point **C** in the graph on the previous page, located in quadrant II.
3. To graph the point (–1,–4), start at the origin. Go to the left 1 unit and from there go down 4 units. This is point **D** in the graph on the previous page, located in quadrant III.
4. To graph the point (5,–3), start at the origin. Go to the right 5 units and from there go down 3 units. This is point **E** in the graph on the previous page, located in quadrant IV.

## ▶ Formulas Related to Coordinate Geometry

GLOSSARY

**SLOPE** the steepness of a line
**MIDPOINT** the location halfway between any two points
**DISTANCE** the length of any line segment calculated from the two endpoints

The *slope* between two points $(x_1, y_1)$ and $(x_2, y_2)$ can be found by using the formula $\frac{\text{change in } y}{\text{change in } x} = \frac{y_1 - y_2}{x_1 - x_2}$. Slope is known as *the rise over the run*. In other words, the number in the numerator (top number) tells how many units to move up or down and the number in the denominator (bottom number) tells how many units to move across to the right or left.

 SHORTCUT

Always think of the slope of a line as an improper or proper fraction. A slope of 5 is really a slope of $\frac{5}{1}$ (up 5 units and over 1 unit) and a slope of $1\frac{1}{2}$ is really a slope of $\frac{3}{2}$ (up 3 units and over 2 units).

*Example:*

Find the slope of the line between the points (–2,5) and (3,3).
Use the formula slope $= \frac{\text{change in } y}{\text{change in } x} = \frac{y_1 - y_2}{x_1 - x_2} = \frac{5 - 3}{-2 - 3} = \frac{2}{-5}$. The slope of the line is $-\frac{2}{5}$.

## RULE BOOK

When dealing with *negative slopes*, the negative sign can be written in three different ways: in the numerator, denominator, or in front of the fraction in the middle. $\frac{-1}{2}$, $\frac{1}{-2}$, or $-\frac{1}{2}$. If there are two negative numbers that form this slope, this is actually a positive slope. $\frac{-1}{-2} = \frac{1}{2}$.

To find the *midpoint* between any two points $(x_1, y_1)$ and $(x_2, y_2)$, use the formula

$$\left(\frac{x_1 + x_2}{2}, \frac{y_1 + y_2}{2}\right)$$

*Example:*
Find the midpoint between $(3,4)$ and $(-1,6)$.
Use the formula $\left(\frac{x_1 + x_2}{2}, \frac{y_1 + y_2}{2}\right) = \left(\frac{3 + -1}{2}, \frac{4 + 6}{2}\right) = \left(\frac{2}{2}, \frac{10}{2}\right) = (1, 5)$.
The midpoint is $(1,5)$.

## SHORTCUT

To find the midpoint, you are adding two values of *x* and *y* and then dividing by two, which is how you find the average of two numbers. Think of the midpoint as the average *x*-value and average *y*-value.

To find the *distance* between any two points $(x_1, y_1)$ and $(x_2, y_2)$, use the formula

$$d = \sqrt{(x_1 - x_2)^2 + (y_1 - y_2)^2}$$

*Example:*
Find the distance between the points $(1,1)$ and $(4,5)$.
Use the formula and substitute the values for *x* and *y*.
$\sqrt{(x_1 - x_2)^2 + (y_1 - y_2)^2} = \sqrt{(1 - 4)^2 + (1 - 5)^2} = \sqrt{(-3)^2 + (-4)^2} = \sqrt{9 + 16} = \sqrt{25} = 5$.
The distance between the two points is five units.

## ▶ Graphing Equations

One way to graph any equation is to find points in the plane that satisfy the equation. In other words, choose an *x*-value and substitute that value for *x* in the equation. Solve the equation for *y*, and then use the two numbers $(x, y)$ as a point. This point will represent one location on the line. Making a table of values will give you points to connect to form the line.

GLOSSARY

**LINEAR EQUATION** an equation of two variables ($x$ and $y$) whose graph is always a straight line; $y = 4x + 5$ is a linear equation.

*Example:*

Graph the linear equation $y = 2x - 1$ using a table.

First, construct a table and choose a few $x$-values. You need at least two. Choose numbers that form a simple pattern (e.g., all differ by one) and then you will also see a pattern in the $y$-values. This should help you to complete the table accurately. A good place to start is the set {0, 1, and 2}.

| x | y = 2x − 1 | y |
|---|---|---|
| 0 | | |
| 1 | | |
| 2 | | |

Take each $x$-value and substitute for $x$ in the equation $y = 2x - 1$.

| x | y = 2x − 1 | y |
|---|---|---|
| 0 | 2(0) −1 | −1 |
| 1 | 2(1) −1 | 1 |
| 2 | 2(2) −1 | 3 |

Each pair of $x$- and $y$-values form a point that can be plotted on a coordinate grid to form the line $y = 2x - 1$. This is shown in the figure below.

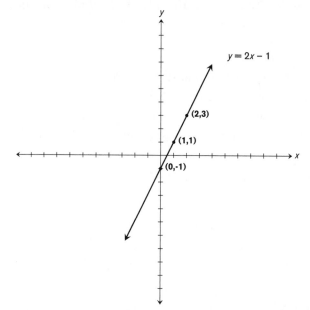

SHORTCUT

A quick way to graph linear equations is to use $y = mx + b$ form.

## ▶ Slope-Intercept Form of Linear Equations

Linear graphing can also be done another way by using the slope-intercept form of the equation. This is also called $y = mx + b$ form, where $m$ represents the slope, or steepness, of the line and $b$ represents the $y$-intercept of the line.

### GLOSSARY

THE *Y*-INTERCEPT of a line is the place where the line crosses the $y$-axis. It will cross at the $b$ value of the equation $y = mx + b$. This point is also known as $(0,b)$.

In order to use this method, first practice getting the equations in the right form. To use this form, make sure your equation has the $y$-value by itself on one side of the equal sign. On the other side of the equation, the number in front of $x$ is the slope of the line and the number being added or subtracted is the $y$-intercept. This procedure is the same as solving literal equations from Chapter 2.

*Examples:*
Find the slope and $y$-intercept of each linear equation by using $y = mx + b$ form.

**a.** $y = \frac{2}{3}x - 1$
This equation is already in the form $y = mx + b$. The number with $x$ is $\frac{2}{3}$, so $\frac{2}{3}$ is the slope of the line. The number on the end is $-1$, so $-1$ is the $y$-intercept.

**b.** $-2y = 4x + 10$
This equation is not in the form $y = mx + b$ but can be rearranged, or transformed, to the correct form. In order to do this, $y$ must be on one side of the equation alone: $-2y = 4x + 10$.
Solve for $y$ by dividing both sides by $-2$: $\frac{-2y}{-2} = \frac{4x}{-2} + \frac{10}{-2}$.
This equation simplifies to $y = -2x - 5$.
The slope of the line is $-2$ or $-\frac{2}{1}$, and the $y$-intercept is $-5$.

 RULE BOOK

Be sure that when you divide or multiply both sides of an equation that you perform the operation on every term of the equation. Note that when dividing $-2y = 4x + 10$ by $-2$, all three terms were divided by $-2$; $\frac{2y}{-2} = \frac{4x}{-2} + \frac{10}{-2}$.

**c.** $-4x + 3y = 12$

This equation is not in the form $y = mx + b$.
Solve for $y$ to get into the correct form: $-4x + 3y = 12$.
Add $4x$ to both sides of the equation: $-4x + 4x + 3y = 12 + 4x$.
Simplify and write the $x$ term first in order to get into $mx + b$ form:
$3y = 4x + 12$.
Divide both sides by 3: $\frac{3y}{3} = \frac{4x}{3} + \frac{12}{3}$.
Simplify: $y = \frac{4}{3}x + 4$.
The slope of the line is $\frac{4}{3}$ and the $y$-intercept is 4.

Once the equation is in $y = mx + b$ form, you are ready to graph. You can use the following page to practice. First, start at the point $(0,b)$ using the $y$-intercept of the equation as $b$. Make a dot at this location on the graph. From there, use the slope of the line. If your slope isn't already a fraction, make it into one. Use the top number as the *rise* of the line, or the number of units to count up from $b$. Then use the denominator as the *run*, or the number of units to count over. Make a dot at this new location. Starting at this new point, repeat this process two or three more times and connect the dots. You should now have a straight line that represents the equation.

 RULE BOOK

When graphing linear equations you should always get a straight line as a result.

Let's practice graphing by using the equations from the previous examples.

**a.** $y = \frac{2}{3}x - 1$

The slope of this line is $\frac{2}{3}$ and the $y$-intercept is $-1$. Start at the point $(0,-1)$ on the $y$-axis and make a dot at this location. Since the slope is $\frac{2}{3}$, count up 2 units from $-1$ and then over 3 units to the right. Make a dot at this location. You are now at the point $(3,1)$. Repeat this process again and you will end up at the point $(6,3)$. Connect the dots and put arrows on the end to show the line continues infinitely in both directions.

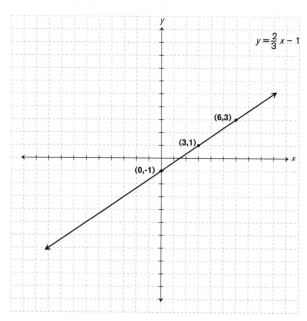

## RULE BOOK

When graphing *positive* slopes, first count *up* and then over to the right to find other points on the line. When graphing *negative* slopes first count *down* and then over to the right to find other points on the line.

## SHORTCUT

**Lines that slant *uphill* have positive slope.**

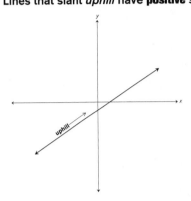

**Lines that slant *downhill* have negative slope.**

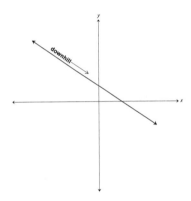

**b.** $-2y = 4x + 10$

This equation transformed to the form $y = -2x - 5$. The slope of this line is $-\frac{2}{1}$ and the $y$-intercept is $-5$. Start at the point $(0,-5)$ on the $y$-axis and make a dot at this location. Since the slope is $-\frac{2}{1}$, count down 2 units from $-5$ and then over 1 unit to the right. Make a dot at this location. You are now at the point $(1,-7)$. Repeat this process again and you will end up at the point $(2, -9)$. Connect the dots and put arrows on the end to show the line continues infinitely in both directions.

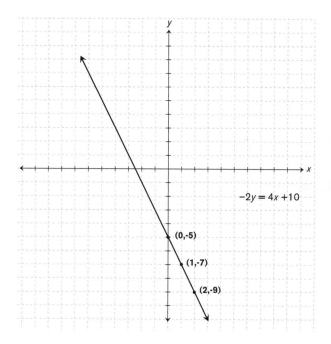

$-2y = 4x +10$

(0,-5)

(1,-7)

(2,-9)

## EXTRA HELP

For practice experimenting with slope and equations of lines, go to the website www.exploremath.com. On the home page under **Gizmos by category**, select **Lines/Linear Equations**. Here you will find numerous lessons and activities on slope calculation and slope-intercept form, as well as many other topics.

**c.** $-4x + 3y = 12$

This equation transformed to the form $y = \frac{4}{3}x + 4$. The slope of this line is $\frac{4}{3}$ and the $y$-intercept is 4. Start at the point $(0,4)$ on the $y$-axis and make a dot at this location. Since the slope is $\frac{4}{3}$, count up 4 units from 4 and then over 3 units to the right. Make a dot at this location. You are now at the point $(3,8)$. Repeat this process again and you will end up at the point $(6,12)$. Graph 2 or 3 more points to get a good picture of the line. Connect the dots and put arrows on the end to show the line continues infinitely in both directions.

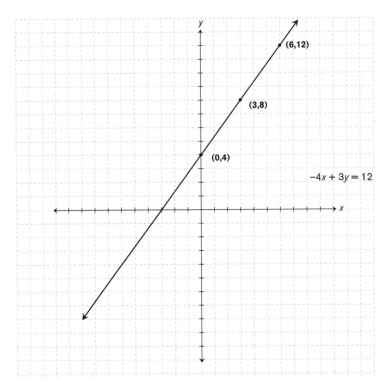

## CALCULATOR TIPS

1. When using a graphing calculator to graph lines they must be in "y =" format. You must have the y alone on one side of the equal sign. Enter your equation into the **Y =** screen and be sure to check your window. Consult your calculator manual for additional questions on graphing equations or setting up a window.

2. When using the **Y =** screen on a graphing calculator, use the **ZOOM** menu as a shortcut to setting up a window for a graph.

### ▶ Two Types of Special Lines

Any equation in the form $y = k$, (where $k$ is a constant) such as $y = 3$ or $y = -2$, will be a horizontal line. The slope of any horizontal line is zero, because there is no change in the $y$-values. In other words, the slope formula will have a zero in the numerator, which makes the entire fraction equal to zero. The equation of the $x$-axis is $y = 0$.

## SHORTCUTS

**1.** To remember that the slope of a horizontal line is zero, think about what it would be like to travel along this line. It would be completely flat so the steepness, or slope, would be zero.

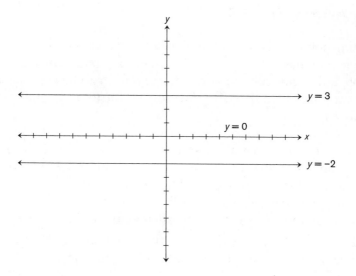

**2.** To remember that the slope of a vertical line is undefined, think about what it would be like to travel along this type of line; probably impossible, or undefined.

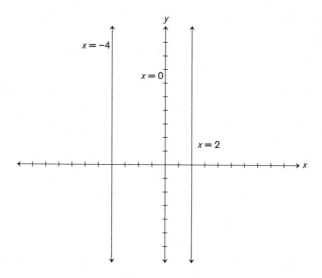

CALCULATOR TIP

Since they cannot be written in the **Y =** screen, use the **DRAW** menu to graph vertical lines in a graphing calculator.

## TIPS AND STRATEGIES

Graphing points and lines in the coordinate plane is not difficult if a few of the basics are mastered. Skills such as plotting points, working with slope, and converting equations to slope-intercept form are the first steps to mastering coordinate geometry. Try focusing on the following tips to help you on your way.

- When graphing points, be sure to always move over to the right or left first, and then up or down. To remember this think $x$ comes before $y$ alphabetically, so find the $x$ value first.
- The slope, or steepness, of a line is the *rise over the run*. Always move up first and over second when graphing using slope.
- There are 3 ways to write a negative slope, such as $\frac{-1}{4}, \frac{1}{-4}, -\frac{1}{4}$.
- Remember that the graph of any linear equation will be a straight line.
- To graph lines, change linear equations to $y = mx + b$ form first and graph them using the slope and the $y$-intercept.
- When graphing a positive slope, count up and then over to the right.
- When graphing a negative slope, count down and then over to the right.
- Lines that slant up to the right have a positive slope.
- Lines that slant up to the left have a negative slope.
- Horizontal lines are in the form $y = k$ (like $y = 3$) and have slope of zero.
- Vertical lines are in the form $x = k$ (like $x = 4$) and have undefined slope, or no slope.
- If a graphing calculator is available and allowed on your exam, use it to graph your equations but remember to put them in $y = mx + b$ form first. Use the **ZOOM** menu to help set up the window.

EXTRA HELP

For more information and practice graphing linear equations, see *Algebra Success in 20 Minutes a Day*, Lesson 8: Graphing Linear Equations.

## CHAPTER QUIZ

Try the following questions for additional practice on coordinate geometry. Check your answers at the end to track your progress and understanding of this topic.

1. In what quadrant is the point (4,–5) located?
   a. I
   b. II
   c. III
   d. IV
   e. none of these

2. Point **P** in the graph below is located at

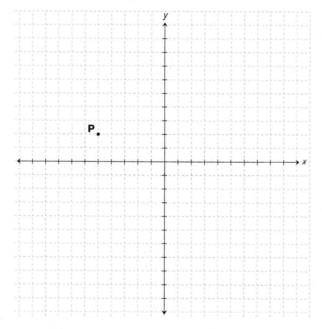

   a. (5,2)
   b. (–5,2)
   c. (2,–5)
   d. (–2,–5)
   e. (–5,–2)

3. What is the slope of a line containing the points (2,–4) and (6,2)?
   a. $\frac{2}{3}$
   b. $-\frac{1}{2}$
   c. 0
   d. $-\frac{3}{4}$
   e. $\frac{3}{2}$

4. What is the length of line segment AB if point A is located at (–1,–5) and point B is located at (4,7)?
   a. 13
   b. 7
   c. $\sqrt{13}$
   d. $\sqrt{29}$
   e. $\sqrt{119}$

5. If the endpoints of a line segment are located at (–4,–4) and (–2,6), what is the midpoint of the segment?
   a. (3,1)
   b. (–3,1)
   c. (3,–1)
   d. (–1,1)
   e. (–1,3)

6. What is the slope of the equation $y = -3x + 9$?
   a. 3
   b. 9
   c. –3
   d. $-\frac{1}{3}$
   e. $\frac{1}{9}$

7. What is the y-intercept of the equation $y = \frac{1}{2}x - 5$?
   a. $-\frac{1}{5}$
   b. –5
   c. $\frac{1}{2}$
   d. 2
   e. 5

8. Which of the following is the equation of a line with a slope of –1 and a y-intercept of 7?
   a. $y = 7x - 1$
   b. $y = -7x - 1$
   c. $y = -x + 7$
   d. $y = x + 7$
   e. $y = x - 7$

9. Which of the following is the equation of a line with a slope of $\frac{2}{3}$ and a $y$-intercept of $-3$?
   a. $2x + 3y = -9$
   b. $-2x + 3y = -9$
   c. $-2x + 3y = 9$
   d. $2x - 3y = 9$
   e. $-2x - 3y = -9$

10. Which of the following represents the equation $-2x + y = 10$ written in slope-intercept form?
    a. $y = -2x + 10$
    b. $y = 2x - 10$
    c. $y = -2x - 10$
    d. $y = 2x + 10$
    e. $y = 10 - 2x$

11. Which of the following represents the equation $6x - 4y = 8$ written in slope-intercept form?
    a. $y = -\frac{3}{2}x - 2$
    b. $y = \frac{3}{2}x - 2$
    c. $y = -\frac{3}{2}x + 8$
    d. $y = \frac{3}{2}x - 8$
    e. $y = \frac{2}{3}x - 2$

**12.** Which of the following is the graph of the equation $y = 2x + 1$?

a.

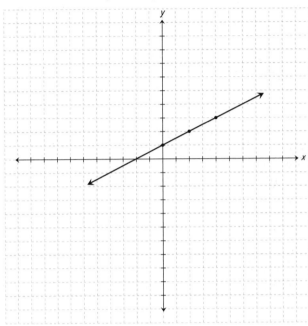

b.

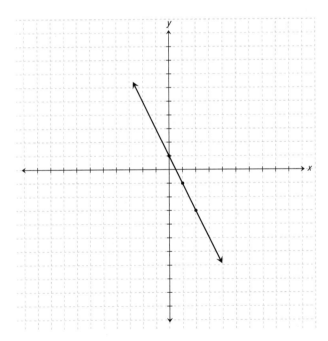

**c.**

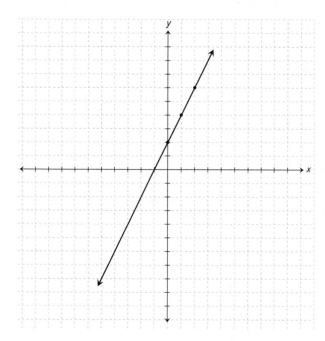

**d.**

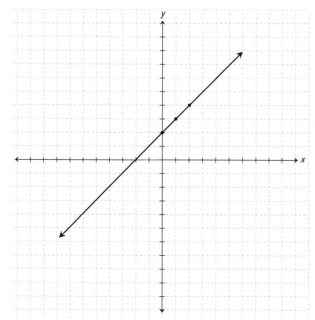

e.

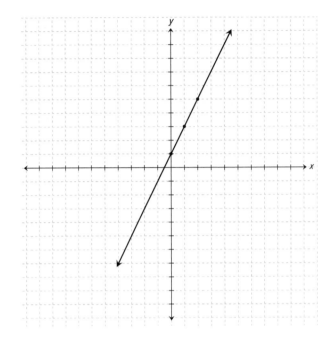

13. Which of the following points is a solution to the graph
$5y - 10 = x$?
a. (2,0)
b. (0,10)
c. (3,5)
d. (5,15)
e. (15,5)

14. For which of the following equations is the point (–6,–3) a
solution?
a. $y = 3x - 6$
b. $y = -6x + 3$
c. $y = 3x$
d. $3y = -6x$
e. $y - 3 = x$

**15.** Which of the following is the graph of the equation $y = x$?

a.

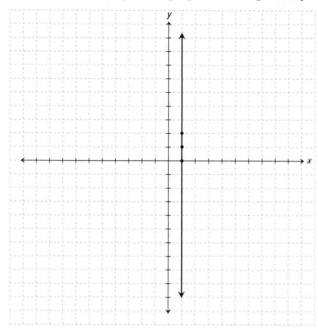

b.

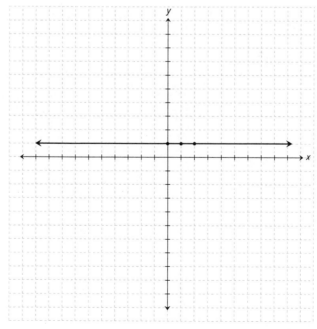

c.

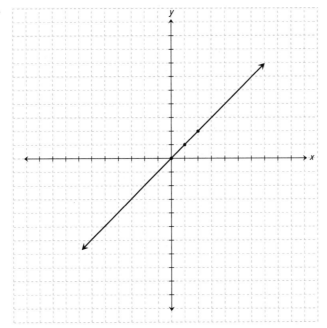

d.

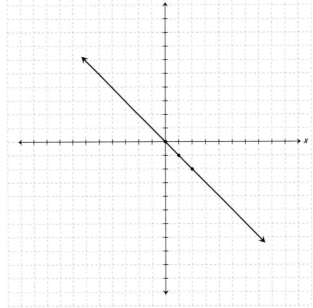

**e.**

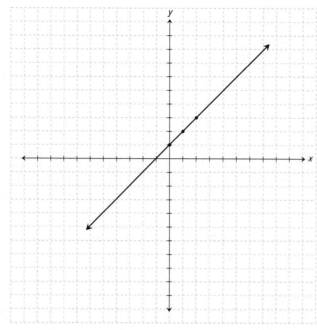

**16.** Which of the following is the graph of the equation $5y - x = 5$?

**a.**

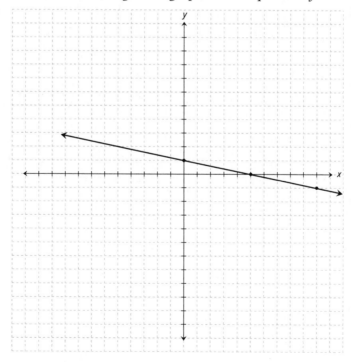

**b.**

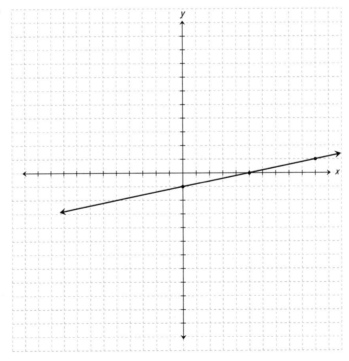

**c.**

**d.**

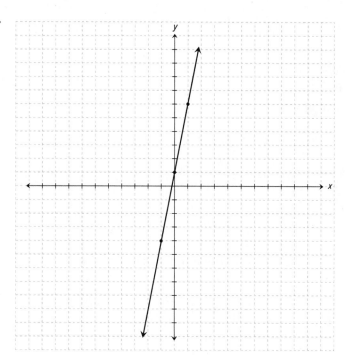

**e.**

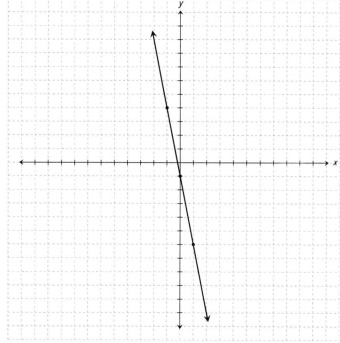

**17.** What is the slope of the line $y = -5$?

    **a.** 0

    **b.** 1

    **c.** $-5$

    **d.** 5

    **e.** undefined

**18.** What is the $y$-intercept of the graph $x = 8$?

    **a.** 0

    **b.** 1

    **c.** $-8$

    **d.** 8

    **e.** none of these

**19.** If the point $(3,n)$ is on the graph of the equation $y = -3x + 7$, what is the value of $n$?

    **a.** $-3$

    **b.** $-2$

    **c.** 2

    **d.** 7

    **e.** 16

**20.** If the point $(m,-1)$ lies on the graph of the equation $x - 6y = 8$, what is the value of $m$?

    **a.** $-6$

    **b.** $-2$

    **c.** 2

    **d.** 6

    **e.** 8

**21.** Which of the following is the graph of the equation $-4x - 2y = 8$?

**a.**

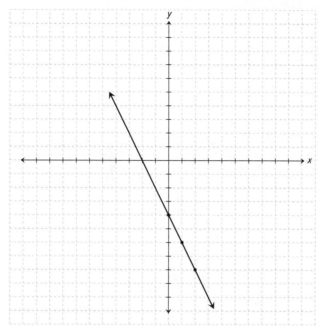

**b.**

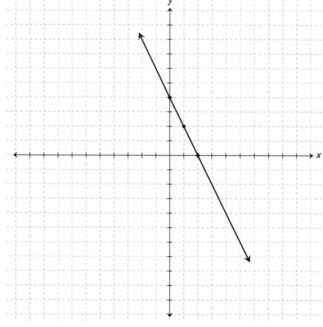

**c.**

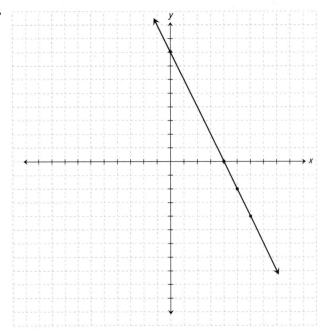

**d.**

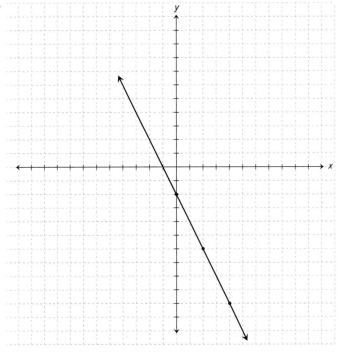

e.

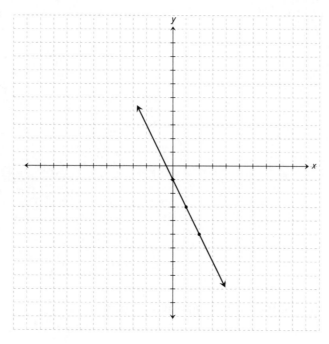

**22.** Which of the following equations matches the *x-y* table?

| x | y |
|---|---|
| −1 | 2 |
| 0 | 4 |
| 1 | 6 |
| 2 | 8 |
| 3 | 10 |

**a.** $y = 2x - 1$
**b.** $3y = x + 2$
**c.** $y = 2x + 4$
**d.** $x - y = 4$
**e.** $x + 3 = y$

**23.** Which of the following graphs has positive slope?

**a.**

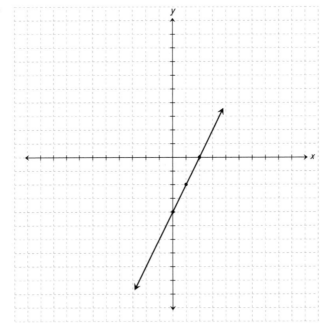

b.

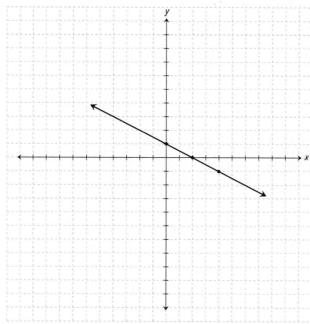

c.

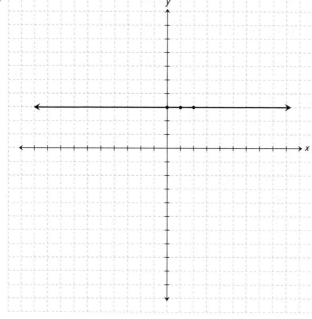

**d.**

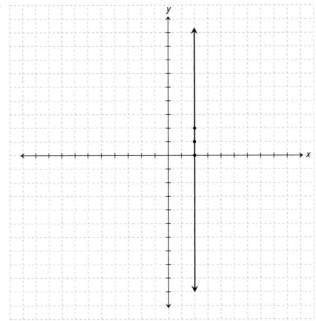

**e.** none of these

**24.** Which table represents a linear equation?

**a.**

| x | y |
|---|---|
| 0 | 0 |
| 1 | 1 |
| 2 | 4 |
| 3 | 9 |

**b.**

| x | y |
|---|---|
| −1 | 0 |
| 1 | 2 |
| 3 | 0 |

**c.**

| x | y |
|---|---|
| −3 | −3 |
| 0 | −2 |
| 3 | −1 |

d.

| x  | y  |
|----|----|
| -2 | -8 |
| -1 | -1 |
| 0  | 0  |
| 2  | 8  |

e.

| x | y |
|---|---|
| 0 | 0 |
| 1 | 1 |
| 4 | 2 |
| 9 | 3 |

25. The slope of a particular linear equation is zero and does not coincide with the x-axis. What can be determined about this line?
    a. The line crosses the y-axis at zero.
    b. The line crosses through quadrants I and III.
    c. The line does not have an x-intercept.
    d. The line can be written in the form x = *a number*.
    e. The line is not a straight line.

## ANSWERS

Here are the answers and explanations for the chapter quiz. Read over the explanations carefully to correct any misunderstandings. Refer to Learning-Express's *Algebra Success in 20 Minutes a Day* for further review and practice.

1. d. To find the location of the point (4,–5) start at the origin and move 4 units to the right. From that point, move 5 units down. Moving over to the right and down puts the point in quadrant IV.

2. b. To get to point **P**, start at the origin and move 5 units to the left. Moving to the left is a negative direction, so the x-coordinate is –5. From there, move 2 units up. Moving up is a positive direction, so the y-coordinate is 2. Point **P** is located at (–5,2).

**3. e.** Use the slope formula. Slope = $\frac{\text{change in } y}{\text{change in } x} = \frac{y_1 - y_2}{x_1 - x_2} = \frac{-4 - 2}{2 - 6} = \frac{-6}{-4} = \frac{3}{2}$.
The slope of the line is $\frac{3}{2}$.

**4. a.** Use the formula and substitute the values for $x$ and $y$;
$\sqrt{(x_1 - x_2)^2 + (y_1 - y_2)^2} = \sqrt{(-1 - 4)^2 + (-5 - 7)^2} =$
$\sqrt{(-5)^2 + (-12)^2} = \sqrt{25 + 144} = \sqrt{169} = 13$.
The distance between the two points is 13 units.

**5. b.** Use the formula $\left(\frac{x_1 + x_2}{2}, \frac{y_1 + y_2}{2}\right) = \left(\frac{-4 + -2}{2}, \frac{-4 + 6}{2}\right) = \left(\frac{-6}{2}, \frac{2}{2}\right) = (-3,1)$. The
midpoint is $(-3,1)$.

**6. c.** The slope $m$ of a linear equation in the form $y = mx + b$ is the
number, or coefficient, in front of $x$. In this equation, the slope
is $-3$.

**7. b.** The $y$-intercept of a linear equation in the form $y = mx + b$ is the
value $b$. The equation $y = \frac{1}{2}x - 5$ is in this form; $b$ is $-5$.

**8. c.** Using the form $y = mx + b$, $m$ represents the slope of the line and $b$
represents the $y$-intercept. In this problem, $m = -1$ and $b = 7$. An
equation using these values would be $y = -1x + 7$, which is also
equal to $y = -x + 7$.

**9. b.** The equation of a line with a slope of $\frac{2}{3}$ and a $y$-intercept of $-3$ is $y$
$= \frac{2}{3}x - 3$. Since this is not an answer choice, change the equation
around to see if it matches any of the ones listed. Subtract $\frac{2}{3}x$ from
both sides to get the $x$ and $y$ on the same side of the equal sign;
$y - \frac{2}{3}x = \frac{2}{3}x - \frac{2}{3}x - 3$. This simplifies to $y - \frac{2}{3}x = -3$. Multiply each side
of the equation by 3 to eliminate the fraction; $3y - 3(\frac{2}{3}x) = -3(3)$.
This simplifies to $3y - 2x = -9$ which is equivalent to $-2x + 3y = -9$.

**10. d.** The slope-intercept form of an equation is written as $y = mx + b$,
where $m$ is the slope of the line and $b$ is the $y$-intercept. To change
the equation to this form, get $y$ alone by adding $2x$ to both sides of
the equation; $-2x + 2x + y = 10 + 2x$. Simplify and write the $x$ term
first on the right side; $y = 2x + 10$. The correct answer choice is **d**.

**11. b.** The slope-intercept form of an equation is written as $y = mx + b$,
where $m$ is the slope of the line and $b$ is the $y$-intercept. To change
the equation to this form, get $y$ alone by subtracting $6x$ from both

sides of the equation; $6x - 6x - 4y = 8 - 6x$. Simplify and write the $x$ term first on the right side of the equation; $-4y = -6x + 8$. Divide each side of the equation by $-4$; $\frac{-4y}{-4} = \frac{-6x}{-4} + \frac{8}{-4}$. This simplifies to $y = \frac{3}{2}x - 2$, which is choice **b**.

12. **e.** The equation of the line $y = 2x + 1$ has a slope of 2 and a $y$-intercept of 1. First look for the answer choices that cross the $y$-axis at (0,1). This occurs in choices **a**, **b**, and **e**. Out of those three choices, look for graphs with positive slope, or lines that go up to the right. This occurs in choices **a** and **e**. Since a slope of 2 is equal to $\frac{2}{1}$, examine choices **a** and **e** to find the line that has a change in $y$ of 2 (goes up 2 units) and a change in $x$ of 1 (goes over one unit). This is choice **e**. Choice **a** goes up 1 and over 2 which is a slope of $\frac{1}{2}$.

13. **e.** To solve this type of question is to substitute the values for $x$ and $y$ into the equation to see which choice makes the equation true. Substitute (2,0) to get $5(0) - 10 = 2$. This results in $0 - 10 = 2$ which simplifies to $-10 = 2$. This is not true. Substitute (0, 10) to get $5(10) - 10 = 0$. This results in $50 - 10 = 0$ which simplifies to $40 = 0$. This is also not true. Substitute (3,5) to get $5(5) - 10 = 3$. This results in $25 - 10 = 3$ which simplifies to $15 = 3$. This is also not true. Substitute (5,15) to get $5(15) - 10 = 5$. This results in $75 - 10 = 5$ which simplifies to $65 = 5$. At this point we assume that **e** is the correct answer. Substitute (15,5) to get $5(5) - 10 = 15$. This results in $25 - 10 = 15$ which simplifies to $15 = 15$ which is true.

14. **e.** Take the point $(-6,-3)$ and substitute $x = -6$ and $y = -3$ into each equation. For choice **a**, the equation becomes $-3 = 3(-6) - 6$. This simplifies to $-3 = -18 - 6$ which is equal to $-3 = -24$. This is not true. For choice **b**, the equation becomes $-3 = -6(-6) + 3$. This simplifies to $-3 = 36 + 3$ which is equal to $-3 = 39$. This is also not true. For choice **c**, the equation becomes $-3 = 3(-6)$. This simplifies to $-3 = -18$ which is not true. For choice **d**, the equation becomes $3(-3) = -6(-6)$. This simplifies to $-9 = 36$ which is also not true. For choice **e**, the equation becomes $-3 - 3 = -6$. This simplifies to $-6 = -6$ which is true.

15. **c.** The equation $y = x$ has a slope of 1 and a $y$-intercept of 0. First, look for the answer choices that intersect the $y$-axis at (0,0) which is the origin. This occurs in choices **c** and **d**. Of these two choices, find the one with the slope of positive 1. This is choice **c**. In this graph, the line goes up 1 unit and over 1 unit to the right to get from one point to another. Choice **d** has a slope of $-1$.

**16. b.** First change the equation into $y = mx + b$ form by adding $x$ to both sides of the equal sign; $5y - x + x = 5 + x$. Divide both sides by 5 and write the $x$ term first on the right side; $\frac{5y}{5} = \frac{x}{5} + \frac{5}{5}$. This simplifies to $y = \frac{1}{5}x + 1$ which is a line with a slope of $\frac{1}{5}$ and a $y$-intercept of 1. Remember that a slope of $\frac{1}{5}$ means start at the $y$-intercept of 1 and go up 1 unit and over 5 units to get to the next point. Choice **b** crosses the $y$-axis at 1 and has a slope of $\frac{1}{5}$.

**17. a.** The line $y = -5$ is a horizontal line through $(0,5)$. The slope of any horizontal line is zero.

**18. e.** The line $x = 8$ is a vertical line through $(8,0)$. Since it is a vertical line, it does not cross the $y$-axis. There is no $y$-intercept.

**19. b.** In the point $(3,n)$, 3 is the $x$ value. Substitute 3 for $x$ in the equation and solve for $y$; $y = -3(3) + 7$ becomes $y = -9 + 7$; $y = -2$ so the value of $n$ is $-2$.

**20. c.** In the point $(m,-1)$, $-1$ is the $y$ value. Substitute $-1$ for $y$ in the equation and solve for $x$; $x - 6(-1) = 8$ becomes $x + 6 = 8$. Since $x = 2$ the value of $m$ is 2.

**21. a.** First, change the equation $-4x - 2y = 8$ to slope-intercept form, or $y = mx + b$. Add $4x$ to both sides of the equation; $-4x + 4x - 2y = 8 + 4x$. Simplify and write the $x$ term first on the right side ($mx + b$ form); $-2y = 4x + 8$. Divide each side of the equation by $-2$; $\frac{-2y}{-2} = \frac{4x}{-2} + \frac{8}{-2}$. This simplifies to $y = -2x - 4$. For this equation, $m$ (the slope) is equal to $-2$ and $b$ (the $y$-intercept) is equal to $-4$. The correct answer choice will cross the $y$-axis at $-4$ and from there graph each point by moving down 2 units and over 1 unit. The graph that satisfies this is choice **a**.

**22. c.** To find the equation for the table, look for the relationship between the $x$ and $y$ columns and write an equation to express this relationship. In this table, each $y$-value is the result of multiplying the $x$-value by 2 and adding 4. As an equation this is written as $y = 2x + 4$. For example, $y = 2(-1) + 4 = -2 + 4 = 2$. An $x$ value of $-1$ results in a $y$-value of 2. Another example is $y = 2(1) + 4 = 2 + 4 = 6$. An $x$-value of 1 results in a $y$-value of 6.

**23. a.** A line with a positive slope goes up to the right. This is choice **a**. Choice **b** has a negative slope, choice **c** has a slope of zero, and choice **d** has undefined slope.

**24. c.** To find the table that represents a linear equation, graph the points and find the table that makes a straight line. Choice **a** does not make a straight line; it is part of the graph of $y = x^2$.

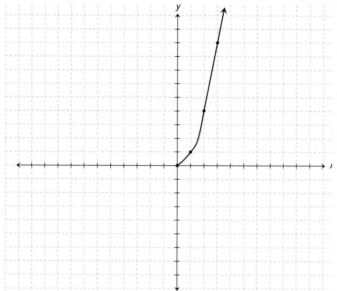

Choice **b** also does not make a straight line when graphed and does not represent a linear equation.

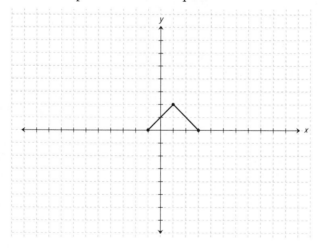

Choice **c** does make a straight line when graphed. The relationship between $x$ and $y$ is the equation $y = \frac{1}{3}x - 2$ which is a linear equation with a slope of $\frac{1}{3}$ and a $y$-intercept of $-2$.

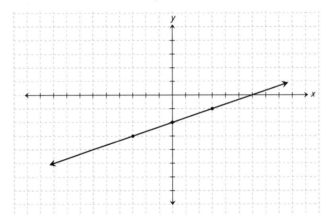

Choice **d** does not make a straight line. These are points from the graph of $y = x^3$.

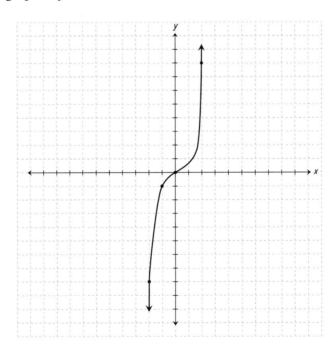

Choice **e** does not make a straight line and is part of the graph of $y = \sqrt{x}$.

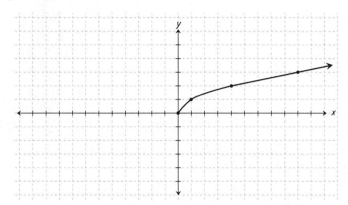

25. **c.** The equation described in this problem has a slope of zero so it is a horizontal straight line in the form of $y = a$ *number*. This eliminates choices **d** and **e**. Since it does not coincide with the $x$–axis, it is either above the $x$-axis or below it, and does not cross it. This line does not have an $x$-intercept. You can conclude that choice **c** is true and this eliminates choice **a**. There is not enough information to tell if the graph is located in quadrants I and III, which was choice **b**.

# Systems of Equations

**B**efore you begin learning and reviewing how to solve systems of equations, take a few minutes to take this ten-question *Benchmark Quiz*. These questions are similar to the type of questions that you will find on important tests. When you are finished, check the answer key carefully to assess your results. Your Benchmark Quiz analysis will help you determine how much time you need to spend on this chapter and the specific areas in which you need the most careful review and practice.

## BENCHMARK QUIZ

1. Which of the following is the solution to the system shown on the graph?

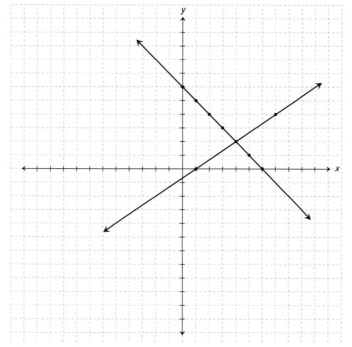

   **a.** (0,6)
   **b.** (1,0)
   **c.** (2,4)
   **d.** (4,2)
   **e.** (6,0)

2. Which of the following choices represents this system of equations graphically?

   $y = x + 1$
   $y = -x + 2$

a.

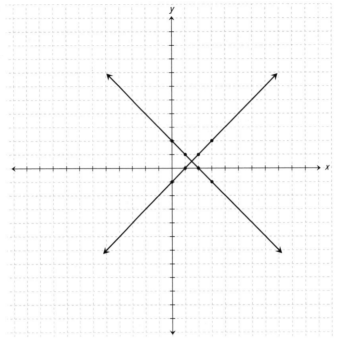

b.

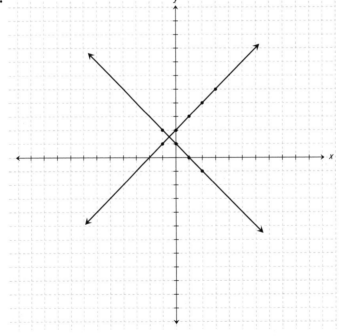

**c.**

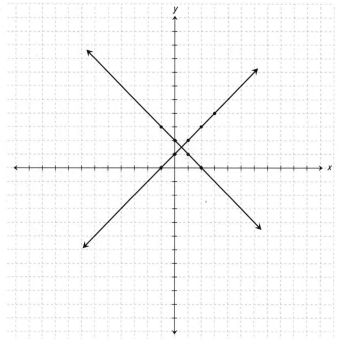

**d.**

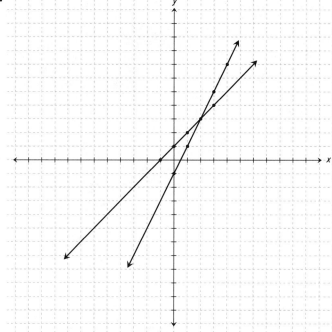

**e.**

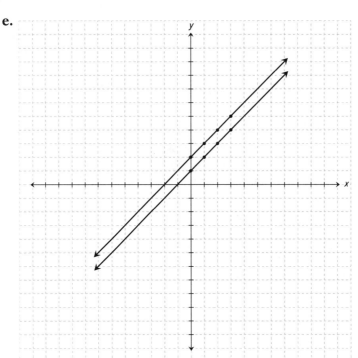

**3.** How many solutions are there in the system shown in the graph?

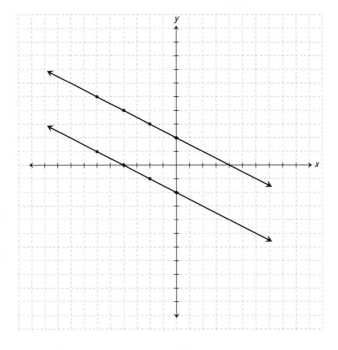

**a.** 0
**b.** 1
**c.** 2
**d.** 3
**e.** infinite solutions

4. Find the solution to the system of equations graphically.

$y = x + 3$
$x + y = 7$

**a.** (5,2)
**b.** (3,4)
**c.** (0,3)
**d.** (4,3)
**e.** (2,5)

5. When solving the following system of equations by elimination method, what would be the next step to eliminate the variable $x$?

$x + y = 5$
$-x + 2y = 6$

**a.** multiplying both equations by 2
**b.** adding the equations
**c.** subtracting the equations
**d.** multiplying the top equation by $-1$
**e.** changing each equation to $y = mx + b$ form

6. Find the value of $x$ in the solution to the system of equations algebraically by elimination.

$-2x + 2y = 24$
$3x + 2y = 4$

**a.** $-8$
**b.** $-4$
**c.** 4
**d.** 6
**e.** 8

7. Find the value of $y$ in the solution to the system of equations algebraically by substitution.

$x = 10 - y$
$x - y = 12$

a. $-1$
b. $0$
c. $1$
d. $2$
e. $11$

8. The sum of two integers is 35 and their difference is 13. Using a system of equations, what is the smaller integer?
a. $-11$
b. $11$
c. $13$
d. $24$
e. $48$

9. Which of the following systems represent two distinct, parallel lines?
a. $x + y = 5$
   $x - y = 4$
b. $2x - 5 = y$
   $-2x + 6 = y$
c. $3x + 2y = 6$
   $y = -\frac{3}{2}x - 8$
d. $y = \frac{1}{2}x + 7$
   $y = -2x - 1$
e. $x = y - 9$
   $y = x + 9$

10. How many solutions are there to the following system of equations?

$2y - 8 = 4x$
$y - 2x = 4$

a. $0$
b. $1$
c. $2$
d. infinite solutions
e. cannot be determined

## BENCHMARK QUIZ SOLUTIONS

How did you do on solving systems of equations? Check your answers here, and then analyze your results to figure out your plan of attack to master these topics.

### ▶ *Answers*

1. **d.** The solution to a system of equations is the point of intersection of the two lines. The point on the graph where the two lines cross is over 4 to the right from the origin and two units up. This is the point (4,2).

2. **c.** In the first equation, you are looking for a line that has a $y$-intercept of 1 and a slope of 1. This line is drawn in both choice **c** and **d**. The second line has a $y$-intercept of 2 and a slope of −1, which only occurs in choice **c**. Since both equations are drawn in choice **c**, this represents the system of equations graphically.

3. **a.** To get from one point to the next on either line you need to go down one unit and over 2. This means that each line has a slope of $-\frac{1}{2}$. Two lines with the same slope are parallel. Parallel lines in the same plane will never intersect; there is no point of solution.

**4. e.** To find the solution graphically, draw both lines on the same set of axes and locate their point of intersection. The first equation has a slope of 1 and a $y$-intercept of 3. The second equation is equal to $y = -x + 7$ in slope-intercept form. This line has a slope of $-1$ and a $y$-intercept of 7.

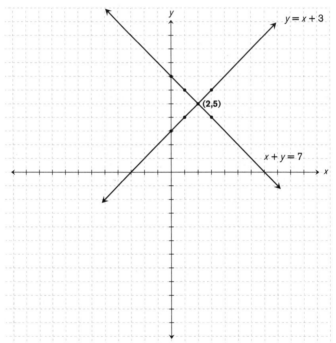

The graphs of the equations intersect at the point (2,5), so this is the solution to the system.

**5. b.** By adding the two equations vertically, $x + -x = 0x = 0$. This process would eliminate $x$ and leave you with $3y = 11$, which would allow you to find the value of $y$ and eventually the solution to the system of equations.

**6. b.** Since you are looking for the value of x, eliminate the variable y. First, multiply the second equation by $-1$ to get the coefficients of y to be opposites. Then add the two equations vertically.

$$-2x + 2y = 24 \quad \Rightarrow \quad -2x + 2y = 24$$
$$-1(3x + 2y = 4) \quad \Rightarrow \quad \underline{-3x - 2y = -4}$$
$$-5x \qquad = 20$$

Since $-5x = 20$, divide each side by $-5$ to get $x = -4$.

**7. a.** Since $x$ is already isolated in the first equation, substitute that value of $x$ (which is $10 - y$) in for $x$ in the second equation; $x - y = 12$ becomes $(10 - y) - y = 12$. Combine like terms on the left side of the equation; $10 - 2y = 12$. Subtract 10 from both sides of the equation. $10 - 10 - 2y = 12 - 10$. This simplifies to $-2y = 2$. Divide both sides by $-2$; $\frac{-2y}{-2} = \frac{2}{-2}$; $y = -1$.

**8. b.** Let $x =$ the larger integer and let $y =$ the smaller integer. Since the sum of the two numbers is 35, add $x$ and $y$ to get the equation $x + y = 35$. Since the difference of the two numbers is 13, subtract $x$ and $y$ to get the equation $x - y = 13$. Combine the two equations vertically to eliminate the variable of $y$.

$$x + y = 35$$
$$\underline{x - y = 13}$$
$$2x \quad\ = 48$$

Since $2x = 48$, divide each side of the equal sign by 2 to get $x = 24$. If $x = 24$, then $24 + y = 35$. So, $y = 11$. To check this result subtract $24 - 11$, which is equal to 13.

**9. c.** Parallel lines have the same slope. To find a system of parallel lines, put the equations in $y = mx + b$ form and look for equations that have the same value of $m$. In choice **a**, $x + y = 5$ becomes $y = -x + 5$ (slope of $-1$) and $x - y = 4$ becomes $y = x - 4$ (slope of 1). These lines are not parallel. In choice **b**, $2x - 5 = y$ has a slope of 2 and $-2x + 6 = y$ has a slope of $-2$. These lines are not parallel. In choice **c**, $3x + 2y = 6$ becomes $y = -\frac{3}{2}x + 3$ (slope of $-\frac{3}{2}$) and $y = -\frac{3}{2}x - 8$ also has a slope of $-\frac{3}{2}$. These lines are parallel. In choice **d**, the slopes of $\frac{1}{2}$ and $-2$ are negative reciprocals. These lines are perpendicular (meet at right angles) but are not parallel. In choice **e**, $x = y - 9$ becomes $y = x + 9$ which is the same as the second equation. Since both equations represent the same line, they are not distinct (different).

**10. d.** Change each equation to $y = mx + b$ form. In the first equation, add 8 to both sides of the equation; $2y - 8 + 8 = 4x + 8$. This simplifies to $2y = 4x + 8$. Divide both sides of the equation by 2; $\frac{2y}{2} = \frac{4x}{2} + \frac{8}{2}$ simplifies to $y = 2x + 4$. In the second equation, add $2x$ to both sides; $y - 2x + 2x = 4 + 2x$. Simplify and write the $x$ term first on the right side of the equation; $y = 2x + 4$. Both equations simplify to $y = 2x + 4$. Since they represent the same line on a graph, every point on the line is a solution. There are infinite solutions to this system of equations.

## BENCHMARK QUIZ RESULTS

If you answered 8–10 questions correctly, you have a good understanding of the in's and out's of solving systems of equations. After reading through the lesson and focusing on the areas you need to review, try the quiz at the end of the chapter to ensure that all of the concepts are clear.

If you answered 4–7 questions correctly, you need to refresh yourself on some of the material. Read through the chapter carefully for review and skill building, and pay careful attention to the sidebars that refer you to more in-depth practice, hints, and shortcuts. Work through the quiz at the end of the chapter to check your progress.

If you answered 1–3 questions correctly, you need help and clarification on the topics in this section. First, carefully read this chapter and concentrate on these skills for solving systems of linear equations. Perhaps you learned this information and forgot it, so take the time now to refresh your skills and improve your knowledge. After taking the quiz at the end of the chapter, you may want to reference a more in-depth and comprehensive book, such as LearningExpress's *Algebra Success in 20 Minutes a Day*.

## JUST IN TIME LESSON—SYSTEMS OF EQUATIONS

In this section, you will learn how to:

- solve a system of equations by graphing
- solve a system of equations algebraically
- recognize three special types of systems
- check your solution to a system

Many of the skills used in this chapter you have already learned and reviewed in previous sections. This chapter will apply those techniques to solving and checking different systems of equations.

GLOSSARY
**SYSTEM OF EQUATIONS** a group of two or more equations

### ▶ *Solving Systems of Equations Graphically*

In order to solve a system of equations by graphing, each equation is graphed and then the point of intersection of the lines is identified. This point is the place where the two or more equations are equal to each other and is the solution to the system of equations.

*Example:*
Solve the following system of equations by graphing.

$$y = 2x + 5$$
$$x + y = 2$$

First, make sure that each equation is in slope-intercept ($y = mx + b$) form. This technique was introduced in Chapter 3. The first equation is already in the correct form. In the equation $y = 2x + 5$, the slope ($m$) is 2 and the $y$-intercept ($b$) is 5.

The second equation needs to be transposed to the correct form. To do this, subtract $x$ from both sides of the equation; $x + y = 2 \Rightarrow x - x + y = 2 - x$.

Write the $x$-term first on the right side of the equation; $y = -x + 2$.

This equation is now in slope-intercept form. The slope of the line is $-1$ and the $y$-intercept is 2.

To solve this system of equations graphically, graph both equations on the same set of axes and look for where the two lines cross each other, or intersect.

This graph shows the system of $y = 2x + 5$ and $x + y = 2$ on a graph.

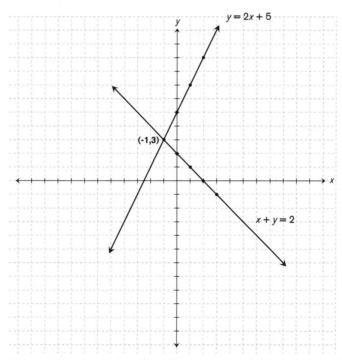

Since the two lines cross at the point ($-1, 3$), the solution to the system is at the point where $x = -1$ and $y = 3$.

 RULE BOOK

The solution to a system of equations on a graph is the *point of intersection* of the two lines.

To check the solution to the system, substitute the $x$ and $y$ value of the solution point into *both* equations.

$$y = 2x + 5 \qquad x + y = 2$$
$$3 = 2(-1) + 5 \qquad -1 + 3 = 2$$
$$3 = 3 \qquad\qquad 2 = 2$$

Since both equations are equal, the solution is correct.

 CALCULATOR TIP

To solve a system of equations in a graphing calculator, enter the equations into the **Y =** screen. Set up an appropriate window and graph the lines. Use the **CALC** menu to calculate the point of intersection.

## ▶ *Three Special Cases of Systems*

There are three special cases of linear equations that you may encounter. The lines may be *parallel*, *perpendicular*, or they may *coincide* with each other. The first case is parallel lines.

### ●●●●GLOSSARY

**PARALLEL LINES** lines in the same plane that have the same slope and will never intersect

Two parallel lines have the same slope. When solving graphically, you may notice that the values for $m$ are the same, even before you graph the equations.

*Example:*
Solve the following system of equations by graphing.
$$x + y = 4$$
$$y = -x - 3$$
First, put each equation into slope-intercept form.
Subtract $x$ from both sides of the first equation and write the $x$ term first on the right side.
$$x - x + y = -x + 4$$
$$y = -x + 4$$
The slope is $-1$ and the $y$-intercept is 4.

The second equation is already in the correct form.

$y = -x - 3$

The slope is –1 and the y-intercept is –3.

Notice that both equations have a slope of –1 but each has a different y-intercept.

Graph both equations on the same set of axes.

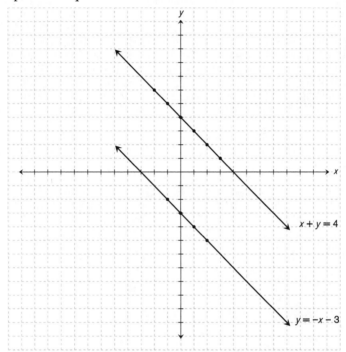

These two graphs run next to each other, or are parallel, but will never cross because the slope of each line is the same. There is no solution to this type of system.

The second case is perpendicular lines.

## GLOSSARY

**PERPENDICULAR LINES** lines that meet to form right angles. Their slopes are negative reciprocals of each other.

**NEGATIVE RECIPROCALS** two fractions where one is positive and the other is negative, and the numerators and denominators are switched. Some pairs of negative reciprocals are $\frac{1}{4}$ and $-4$, $-\frac{5}{6}$ and $\frac{6}{5}$, and 1 and –1.

*Example:*

Solve the following system of equations by graphing.

$y = \frac{3}{2}x - 1$

$y = -\frac{2}{3}x - 1$

Both equations are in the slope-intercept form.
In the first equation, the slope is $\frac{3}{2}$ and the $y$-intercept is –1.
In the second equation, the slope is $-\frac{2}{3}$ and the $y$-intercept is –1 .
Notice that the slopes are negative reciprocals of one another, $\frac{3}{2}$ and $-\frac{2}{3}$.
Graph the two equations on the same set of axes to find the solution.

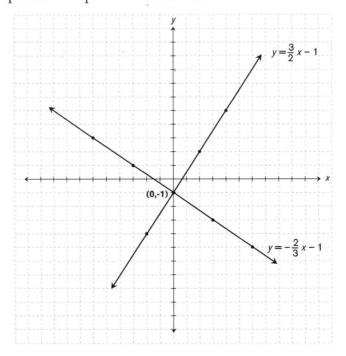

The two lines intersect at the point (0,–1) so this is the solution to the system. These two lines meet to form right angles and will have exactly one solution.

The third type of special case are lines that coincide with one another. *Coincident lines* are two lines that have the same equation. On a graph they represent the same line. The solution set to this type of system is *all* of the points on the line, or infinite solutions.

### ●●●●GLOSSARY

**COINCIDENT LINES** two lines that have the same equation; on a graph they represent the same line

When solving a system of equations that coincide, you will notice something unique as soon as you begin to graph the lines. Each of the equations, although they may appear different at the start, represents the same line.

*Example:*

Solve the following system of equations by graphing.

$$y - 5 = 3x$$
$$2y - 6x = 10$$

Change each equation to slope-intercept form.

Add 5 to both sides of the first equation; $y - 5 + 5 = 3x + 5$.

This simplifies to $y = 3x + 5$.

The slope of the line is 3 and the $y$-intercept is 5.

For the second equation, add $6x$ to both sides; $2y - 6x + 6x = 10 + 6x$.

This simplifies to $2y = 10 + 6x$.

Divide both sides by 2 and write the $x$ term first on the right side of the equation; $\frac{2y}{2} = \frac{6x}{2} + \frac{10}{2}$.

This simplifies to $y = 3x + 5$. The slope of this line is 3 and the $y$-intercept is 5.

Graph both equations on the same set of axes.

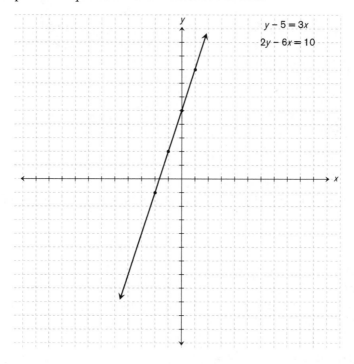

When graphed on a set of axes, both lines appear to be exactly one on top of the other. The graph of each equation is the same line. In this special case, both equations share *all* solutions. Therefore, the solution set to this type of system is *all points on the line* or *infinite solutions*.

RULE BOOK

The solution to a linear system of equations will be either one solution, no solution, or infinite solutions.

 EXTRA HELP

For more information and practice solving systems of equations graphically, see *Algebra Success in 20 Minutes a Day*, Lesson 11: **Graphing Systems of Equations and Inequalities.**

## ▶ *Solving Systems of Equations Algebraically*

When solving a system of equations, you are finding the place or places where two or more equations equal each other. Here are two ways to do this algebraically: by elimination and by substitution.

**Elimination Method.** In this method, you will be using addition or subtraction to eliminate one of the two variables so that the equation you are working with has only one variable.

## ●●●● GLOSSARY

**COEFFICIENT** a number, or constant, being multiplied by a variable. For example in the term $3x$, 3 is the coefficient of the variable, $x$.

Solve the system $x - y = 6$ and $2x + 3y = 7$.
Put the equations one above the other, lining up the $x$'s, $y$'s, and the equal sign.

$$x - y = 6$$
$$2x + 3y = 7$$

Multiply the first equation by $-2$ so that the coefficients of $x$ are opposites. This will allow the $x$'s to cancel out in the next step. Make sure that *all* terms are multiplied by $-2$. The second equation remains the same.

$$-2(x - y = 6) \Rightarrow -2x + 2y = -12$$
$$2x + 3y = 7 \Rightarrow 2x + 3y = 7$$

Combine the new equations vertically.

$$-2x + 2y = -12$$
$$\underline{2x + 3y = \phantom{-}7}$$
$$5y = -5$$

Divide both sides by 5.

$$\frac{5y}{5} = \frac{-5}{5}$$
$$y = -1$$

To complete the problem, solve for $x$ by substituting $-1$ for $y$ into one of the original equations.

$x - (-1) = 6$
$x + 1 = 6$
$x + 1 - 1 = 6 - 1$
$x = 5$

The solution to the system is $x = 5$ and $y = -1$, or $(5,-1)$.

## SHORTCUT

**Use the elimination method when dealing with equations that are not in the form $y = mx + b$ already and contain many different coefficients of $x$ and $y$.**

**Substitution Method.** In this method, you will be substituting a quantity for one of the variables to create an equation with only one variable.

## RULE BOOK

**Always use parentheses when substituting a quantity.**

Solve the system $x + 2y = 5$ and $y = -2x + 7$
Substitute the second equation into the first for $y$.

$x + 2(-2x + 7) = 5$

Use distributive property to remove the parentheses.

$x + -4x + 14 = 5$

Combine like terms. Remember $x = 1x$.

$-3x + 14 = 5$

Subtract 14 from both sides and then divide by $-3$.

$-3x + 14 - 14 = 5 - 14$
$\frac{-3x}{-3} = \frac{-9}{-3}$
$x = 3$

To complete the problem, solve for $y$ by substituting 3 for $x$ in one of the original equations.

$y = -2(3) + 7$
$y = -6 + 7$
$y = 1$

The solution to the system is $x = 3$ and $y = 1$, or $(3,1)$.

## SHORTCUT

**Use the substitution method when one of the equations already has either $x$ or $y$ alone.**

## TIPS AND STRATEGIES

Solving systems of equations applies many of the basics that you have learned from equation solving and graphing in earlier chapters. Here are some additional points to keep in mind while tackling this type of question.

- The solution of a system of equations is the point where the two equations are equal. On a graph, this is the point of intersection.
- The solution to a linear system of equations will either be one solution, no solution, or infinite solutions.
- Two lines with the same slope are parallel and will never intersect. There is no solution to this type of system.
- Two lines that have slopes that are negative reciprocals of each other are perpendicular. They meet at a right angle and there will be one solution.
- Lines with the same slope and $y$-intercept are really the same line and will have an infinite number of solutions (all points on the line).
- When solving algebraically, if one variable of one equation is already isolated, *substitution* may be the easier method.
- When solving algebraically and there are many different coefficients in the equations, *elimination* may be the easier method.
- Remember to always line up like terms when using the elimination method.
- Always check both equations with the solution when solving a system. This will ensure that you have found the correct answer.
- Remember that the solution to a system is a point; you need both an $x$ **and** a $y$ value for your answer.

 EXTRA HELP

The website www.exploremath.com has hands-on practice for exploring and solving systems of equations. Click on **Gizmos by Category** on the home page, choose **Systems** and follow directions from there. For even more information and practice solving systems of equations algebraically, see *Algebra Success in 20 Minutes a Day*, Lesson 12: Solving Systems of Equations Algebraically.

## CHAPTER QUIZ

The following problems provide additional practice with systems of equations. Try them and check your progress through this topic using the answer explanations for help and clarification.

1. Which of the following is the solution to the system shown on the graph?

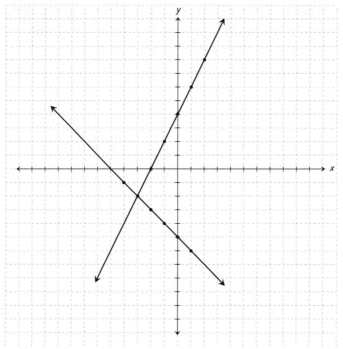

    **a.** (2,3)
    **b.** (–2,–3)
    **c.** (–3,–2)
    **d.** (–2,0)
    **e.** (0,–2)

2. Which of the following is NOT true about a system of linear equations?
    **a.** There could be exactly one solution.
    **b.** There could be exactly two solutions.
    **c.** There could be infinite solutions.
    **d.** There could be three or more equations in the system.
    **e.** There could be no solution.

3. How many solutions are there in a linear system of two distinct parallel lines?
   **a.** 0
   **b.** 1
   **c.** 2
   **d.** 3
   **e.** infinite solutions

4. How many solutions are there to a system of two distinct linear equations where the slopes of the lines are negative reciprocals to each other?
   **a.** 0
   **b.** 1
   **c.** 2
   **d.** infinite solutions
   **e.** cannot be determined

5. By solving the following system of equations by the elimination method, what next step(s) would eliminate the variable $y$?

   $2x - 3y = 12$
   $-x + y = 4$

   **a.** multiplying the second equation by $-2$ and then adding the equations
   **b.** adding the equations, only
   **c.** subtracting the equations, only
   **d.** multiplying the top equation by $-1$ and then adding the equations
   **e.** multiplying the second equation by 3 and then adding the equations

6. Solve the following system of equations by graphing.

   $y = -x + 4$
   $y = x$

   **a.** (0,0)
   **b.** (0,4)
   **c.** (-2,-2)
   **d.** (2,2)
   **e.** (4,0)

7. Which of the following is an equation that would coincide with the line $2y - 4 = 6x$?
   **a.** $3x + y = 2$
   **b.** $2y + 4 = 6x$
   **c.** $y = 3x - 2$
   **d.** $-3x + y = 2$
   **e.** none of these

8. Which of the following shows the system of equations graphically?

   $y = \frac{3}{2}x - 1$
   $-3y = x$

   **a.**

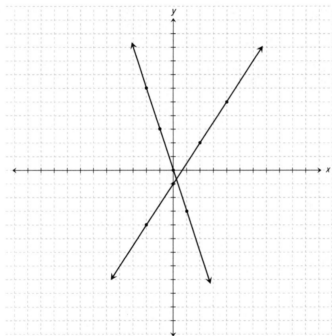

**b.**

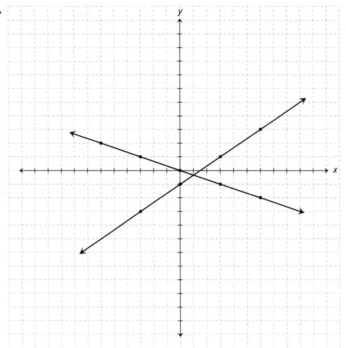

**c.**

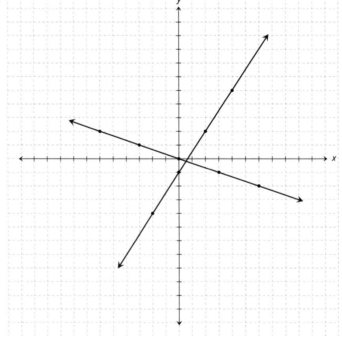

**d.**

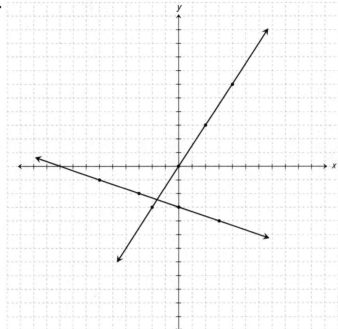

**e.**

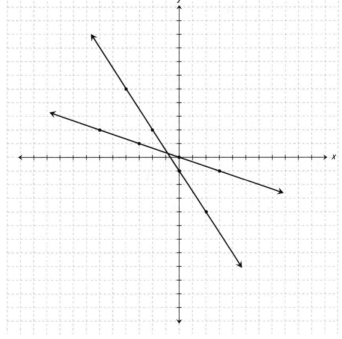

**9.** In which of the following systems is the lines perpendicular?

    **a.** $y = -x - 1$
       $y = 3x - 1$
    **b.** $x + y = 10$
       $y = x + 9$
    **c.** $y - 2x = -3$
       $y - x = -5$
    **d.** $y - x = 3$
       $2y - 2x = 6$
    **e.** $y = x$
       $y = 2x$

**10.** Which graph best represents the system of equations?

$$y - 3 = 2x$$
$$2y + 4x = -6$$

    **a.**

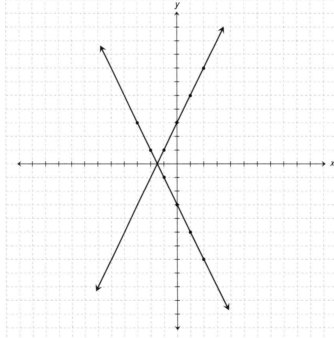

**b.**

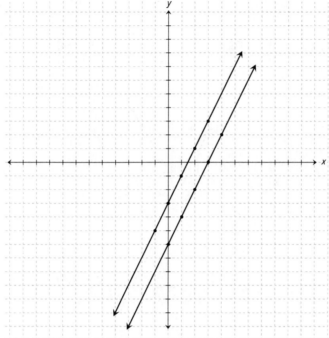

**c.**

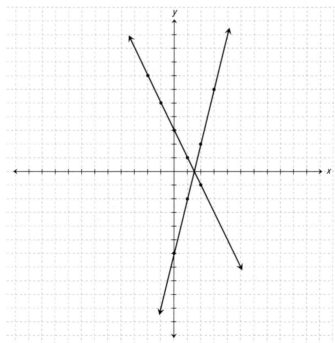

**d.**

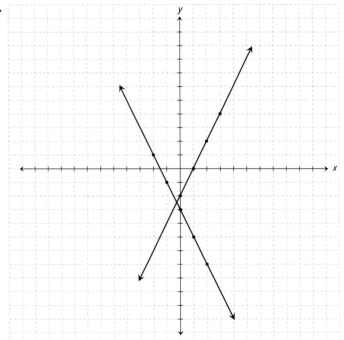

**e.**

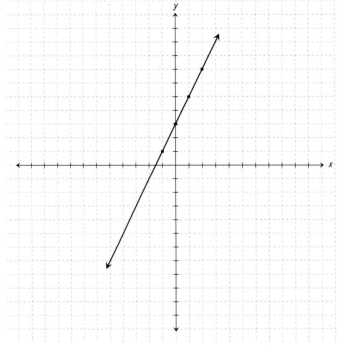

**11.** If the graph of the equation $2x - 3y = 3$ contains the point $(0,-1)$, which of the following graphs also contain this point?
**a.** $2x - 3y = -1$
**b.** $x - y = 3$
**c.** $y = 2x - 3$
**d.** $y = 3x - 1$
**e.** none of these

**12.** What is true of the graphs of $3x = y + 7$ and $y + x = 1$?
**a.** They intersect at $(-2,1)$.
**b.** They do not intersect.
**c.** They represent the same line.
**d.** They intersect at $(2,-1)$.
**e.** They are perpendicular lines.

**13.** Linear equation A has a slope of zero and linear equation B has undefined slope. What can you conclude about the two lines?
**a.** They will intersect at exactly one point.
**b.** They will intersect at exactly two points.
**c.** They will never intersect.
**d.** Both lines have the same $y$-intercept.
**e.** A point of intersection is in quadrant I.

**14.** Which system has infinite solutions?
**a.** $y = 2$
    $5y = 4x + 2$
**b.** $3y = -2x + 6$
    $y = 2x + 2$
**c.** $x = y + 3$
    $x = -y + 5$
**d.** $3x + y = 3$
    $y = -2x - 5$
**e.** none of these

**15.** Find the value of $x$ in the solution to the system of equations algebraically by elimination.

$2x + y = 1$
$3x - 2y = 12$

**a.** $-2$
**b.** $0$
**c.** $2$
**d.** $3$
**e.** $-3$

**16.** Find the value of *y* in the solution to the system of equations algebraically by elimination.

$$3x + 14 = 5y$$
$$2x + 7y = 1$$

**a.** –1
**b.** 1
**c.** –3
**d.** –28
**e.** –31

**17.** Find the value of *x* in the solution to the system of equations algebraically by substitution.

$$y = 7x - 8$$
$$y + 4x = -19$$

**a.** –1
**b.** –8
**c.** –15
**d.** 7
**e.** 15

**18.** Find the value of *y* in the solution to the system of equations algebraically by substitution.

$$-3x + y = 4$$
$$y = 4x + 1$$

**a.** –3
**b.** 3
**c.** 1
**d.** –13
**e.** 13

**19.** Which system of equations has no solution?
   **a.** $y = 2$
      $x = 3$
   **b.** $y = -2x - 1$
      $y = 2x - 5$
   **c.** $y = 0$
      $x = 4 - y$
   **d.** $2x = y$
      $2y = 4x - 6$
   **e.** none of these

**20.** Find the solution to the following system of equations algebraically.

$3x + 4y = 6$
$2x - 6y = 4$

a. (1,2)
b. (4,2)
c. (2,0)
d. (1,4)
e. (6,0)

**21.** What is the solution to the following system of equations?
$-5y + x = 17$ and $x = 2$

a. (0,2)
b. (2,3)
c. (-2,3)
d. (2,-3)
e. (2,0)

**22.** The sum of two integers is 50 and the difference is 18. Use a system of equations to find the larger integer.
a. 16
b. 18
c. 24
d. 34
e. 36

**23.** Four pens and two pencils cost $3.50. Two pens and three pencils cost $2.25. What is the cost of six pens?
a. $.25
b. $.50
c. $3.00
d. $3.50
e. $4.50

**24.** The sum of three times a greater integer and five times the lesser integer is 49. The smaller integer is three less than the greater. What is the value of the lesser integer?
a. -5
b. -3
c. 3
d. 5
e. 8

**25.** A group of 120 people are going camping. They must provide 20 gallons of water for each adult and 15 gallons of water for each child. If the total amount of water they are taking is 1,950 gallons, what is the total number of children in the group?
   **a.** 5
   **b.** 30
   **c.** 45
   **d.** 60
   **e.** 90

## ANSWERS

Here are the answers and explanations for the chapter quiz. Read over the explanations carefully to correct any misunderstandings. Refer to Learning-Express's *Algebra Success in 20 Minutes A Day* Lessons 11–12 for further review and practice.

**1. c.** The solution to the system of equations is the point where the graphs are equal. This is the point of intersection of the two lines. The lines cross at a point that is 3 units to the left of the origin and 2 units down. This is the point $(-3,-2)$.

**2. b.** There could not be exactly two solutions. In a linear system of equations, all lines are straight lines. There could be one solution, infinite solutions, or no solution. In addition, a system of equations is two *or more* related equations, so choice **d** would be true.

**3. a.** There is no solution because the lines will never cross. Two lines that are parallel have the same slope and will never intersect.

**4. b.** This type of system will have one solution. Because the slopes are negative reciprocals of one another, they will meet to form right angles in one location. This point of intersection is the solution to the system.

**5. e.** Eliminate $y$ by getting coefficients of $y$ that are opposites, like $-3$ and 3. Multiplying the second equation by 3 will change it to $-3x + 3y = 12$. Adding the two equations will now eliminate the $y$ terms: $-3y + 3y = 0y = 0$.

**6. d.** In the equation $y = -x + 4$, the slope is $-1$ and the $y$-intercept is 4. In the equation $y = x$, the slope is 1 and the $y$-intercept is 0. The figure shows the graph of the equations on the same set of axes.

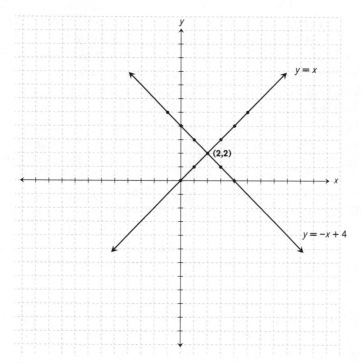

The two lines intersect at the point (2,2). This is the solution to the system of equations.

**7. d.** For two equations to coincide they need to be transformed into the same equation. Choice **b** can be eliminated first because it is in the same form as the original equation but is different because of the subtraction sign. Add 4 to both sides of the original equation to get $2y = 6x + 4$. Then divide each side by 2. The result is $y = 3x + 2$. This eliminates choice **c**. In choice **d** subtract $3x$ from both sides to get $y - 3x = 2$. This is the same as $-3x + y = 2$. This makes the correct answer choice **d** and eliminates choice **a**.

**8. c.** Write each equation in slope-intercept form and graph. The first equation is already in the correct form. The equation $y = \frac{3}{2}x - 1$ has a slope of $\frac{3}{2}$ and a $y$-intercept of $-1$. Of the answer choices, only **a** and **c** have this equation drawn. The equation $-3y = x$ is equal to $y = -\frac{1}{3}x$ in slope-intercept form. The slope of this line is $-\frac{1}{3}$ and the $y$-intercept is 0. Both equations are drawn in choice **c**.

**9. b.** Perpendicular lines meet to form right angles and their slopes are negative reciprocals. To find the two equations whose slopes are negative reciprocals, change each equation to slope-intercept form. Choice **a** is not correct because the first equation has a slope of $-1$ and the second equation has a slope of 3. Choice **b** is correct because the first equation is equal to $y = -x + 10$ which has a slope of $-1$ and the second equation has a slope of 1; $-1$ and 1 are negative reciprocals of one another. Choice **c** is not correct because the first equation is equal to $y = 2x - 3$ which has a slope of 2 and the second equation is equal to $y = x - 5$ which has a slope of 1. Choice **d** is not correct because the first equation is equal to $y = x + 3$ which has a slope of 1 and the second equation is equal to $y = x + 3$ which also has a slope of 1. These equations are actually the same line, or coincident. In choice **e**, the first equation has a slope of 1 and the second has a slope of 2; this would not cause the lines to be perpendicular.

**10. a.** Take both equations and change them to slope-intercept form. Add 3 to both sides of the first equation; $y - 3 + 3 = 2x + 3$. This simplifies to $y = 2x + 3$, a line with a slope of 2 and a $y$-intercept of 3. In the second equation, subtract $4x$ from both sides of the equal sign; $2y + 4x - 4x = -6 - 4x$. Simplify and write the $x$ term first on the right side; $2y = -4x - 6$. Divide both sides by 2; $\frac{2y}{2} = \frac{-4x}{2} - \frac{6}{2}$. This simplifies to $y = -2x - 3$, which has a slope of $-2$ and a $y$-intercept of $-3$. The graph in choice **a** correctly represents both equations.

**11. d.** To find another graph where the point $(0, -1)$ is also a solution, substitute in the values for $x = 0$ and $y = -1$ into each equation as in checking an equation. In choice **a** the equation becomes $2(0) - 3(-1) = -1$. This simplifies to $0 + 3 = -1$. Three does not equal $-1$. In choice **b** the equation becomes $0 - (-1) = 3$. This simplifies to $1 = 3$, which is also not true. In choice **c** the equation becomes $-1 = 2(0) - 3$. This simplifies to $-1 = 0 - 3$. $-1$ does not equal $-3$. In choice **d** the equation becomes $-1 = 3(0) - 1$. This simplifies to $-1 = 0 - 1$ which is equal to $-1 = -1$. This equation is true.

**12. d.** Change the equations to slope-intercept form; $3x = y + 7$ becomes $3x - 7 = y$ by subtracting 7 from both sides of the equal sign. This is equal to $y = 3x - 7$, a line with a slope of 3 and a $y$-intercept of $-7$. Subtract $x$ from both sides of the equation $y + x = 1$ to transform it into $y = -x + 1$. This line has a slope of $-1$ and a $y$-intercept of 1. On a graph the two equations appear as follows.

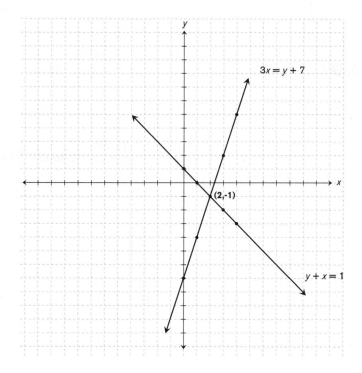

These two lines intersect at the point $(2,-1)$. The answer is choice **d.**

**13. a.** These lines will cross at one point. A line with a slope of zero is a horizontal line and a line with undefined slope is a vertical line. Any two straight lines with different slopes will eventually cross and only cross once. There is not enough information to conclude that either choice **d** or **e** is correct.

**14. e.** To find the system with infinite solutions, look for two coincident lines or lines with the same equation. Choice **a** is eliminated because there is no $x$-term in the first equation and there are both $x$- and $y$-terms in the second equation. In choice **b**, divide the first equation by 3 to get $y = -\frac{2}{3}x + 2$. This is not the same as the second equation. In choice **c** the lines are already in the same form and are not equivalent. In choice **d**, subtract $3x$ from both sides of the first equation to get $y = -3x + 3$. This is not the same as the second equation. The answer is choice **e**, none of these.

**15. c** Since you are looking for the value of $x$, eliminate the variable $y$. First, multiply the first equation by 2 to get the coefficients of $y$ to be opposites. Then add the two equations vertically.

$$2(2x + y = 1) \Rightarrow \quad 4x + 2y = \ 2$$
$$3x - 2y = 12 \Rightarrow \quad 3x - 2y = 12$$
$$\overline{\hspace{3.5cm}7x \qquad = 14} \qquad \text{(Recall that } 2y + -2y = 0)$$

Since $7x = 14$, divide each side by 7 to get $x = 2$.

**16. b.** Since you are looking for the value of $y$, eliminate the variable $x$. First, rewrite the first equation as $3x - 5y = -14$ to line up like terms. Multiply the first equation by 2 and the second equation by $-3$ to get the coefficients of $x$ to be opposites. Then add the two equations vertically.

$$2(3x - 5y = -14) \Rightarrow \quad 6x - 10y = -28$$
$$-3(2x + 7y = 1) \quad \Rightarrow \quad \underline{-6x - 21y = -3}$$
$$-31y = -31$$

Since $-31y = -31$, divide each side by $-31$ to get $y = 1$.

**17. a.** Since the variable $y$ is isolated in the first equation, substitute $7x - 8$ for $y$ in the second equation; $y + 4x = -19$ becomes $(7x - 8) + 4x = -19$. Combine like terms on the left side of the equation; $11x - 8 = -19$. Add 8 to both sides; $11x - 8 + 8 = -19 + 8$. This simplifies to $11x = -11$. Divide both sides by 11; $\frac{11x}{11} = \frac{-11}{11}$. $x = -1$.

**18. e.** Since $y$ is isolated in the second equation, substitute $4x + 1$ for $y$ in the first equation; $-3x + y = 4$ becomes $-3x + (4x + 1) = 4$. Combine like terms on the left side of the equation; $x + 1 = 4$. Therefore $x = 3$. To solve for $y$, substitute $x = 3$ into $y = 4x + 1$; $y = 4(3) + 1 = 12 + 1 = 13$.

**19. d.** For a system of linear equations to have no solution, the lines must be parallel. Look for a pair of equations that have the same slope. In choice **a**, the first line is horizontal and the second line is vertical; they will eventually intersect. In choice **b**, the slopes are $-2$ and 2; they will also intersect. In choice **c**, the first equation is a horizontal line on the $x$-axis and the second equation has a slope of $-1$ ($x = 4 - y$ becomes $y = -x + 4$). They will intersect. In choice **d**, the first equation has a slope of 2. The second equation can be divided by 2 and changed to $y = 2x - 3$, which also has a slope of 2. These lines are parallel and will not intersect.

**20. c.** Since there are many different coefficients in both equations, use the elimination method. Multiply the first equation by 3 and the second equation by 2 to get the coefficients of $y$ to be opposites. Then add the two equations vertically.

$$3(3x + 4y = 6) \Rightarrow 9x + 12y = 18$$
$$2(2x - 6y = 4) \Rightarrow \underline{4x - 12y = 8}$$
$$13x \qquad = 26$$

Since $13x = 26$, divide each side by 13 to get $x = 2$. This question asks for the solution to the system, so substitute $x = 2$ into the first equation; $3x + 4y = 6$ becomes $3(2) + 4y = 6$. This simplifies to $6 + 4y = 6$. Subtract 6 from both sides of the equation; $4y = 0$; $y = 0$. The solution is $(2, 0)$.

**21. d.** Since $x = 2$ is one of the equations, substitute 2 in for $x$ in the first equation; $-5y + x = 17$ becomes $-5y + 2 = 17$. Subtract 2 from both sides of the equation; $-5y + 2 - 2 = 17 - 2$. This simplifies to $-5y = 15$. Divide both sides by $-5$; $\frac{-5y}{-5} = \frac{15}{-5}$; $y = -3$. Since $x = 2$ and $y = -3$, the solution to the system is $(2, -3)$.

**22. d.** Let $x =$ the first integer and let $y =$ the second integer. The equation for the sum of the two integers is $x + y = 50$ and the equation for the difference between the two integers is $x - y = 18$. To solve these by the elimination method combine like terms vertically and the variable $y$ cancels out.

$$x + y = 50$$
$$\underline{x - y = 18}$$

This results in: $\qquad 2x \quad = 68, \quad$ so $x = 34$.

Substitute the value of $x$ into the first equation to get $34 + y = 50$. Subtract 34 from both sides of this equation to get an answer of $y = 16$. The larger integer is 34, choice **d.**

**23. e.** Let $x =$ the cost of one pen and let $y =$ the cost of one pencil. The first statement "four pens and two pencils cost \$3.50" translates to the equation $4x + 2y = 3.50$. The second statement "Two pens and three pencils cost \$2.25" translates to the equation $2x + 3y = 2.25$.

Keep the first equation the same: $\qquad 4x + 2y = 3.50$
Multiply the second equation by $-2$: $\underline{-4x + -6y = -4.50}$
Combine the two equations
 vertically to eliminate $x$: $\qquad\qquad -4y = -1.00$
Divide both sides of the equal sign by $-4$: $\qquad y = .25$

Therefore, the cost of one pencil is \$.25. Since the cost of 4 pens and 2 pencils is \$3.50, $2 \times \$.25 = \$.50$; $\$3.50 - \$.50 = \$3.00$. \$3.00 $\div 4 = \$.75$, so each pen is \$.75. The total cost of 6 pens is $.75 \times 6 = \$4.50$.

**24. d.** Let $x$ = the lesser integer and let $y$ = the greater integer. The first sentence in the question gives the equation $3y + 5x = 49$. The second sentence gives the equation $y - 3 = x$.

| | | |
|---|---|---|
| Substitute $y - 3$ for $x$ in the first equation: | $3y + 5(y - 3)$ | $= 49$ |
| Use the distributive property on the left side of the equation: | $3y + 5y - 15$ | $= 49$ |
| Combine like terms on the left side: | $8y - 15$ | $= 49$ |
| Add 15 to both sides of the equation: | $8y - 15 + 15$ | $= 49 + 15$ |
| Divide both sides of the equation by 8: | $\frac{8y}{8}$ | $= \frac{64}{8}$ |

This gives a solution of $y = 8$. Therefore the lesser, $x$, is three less than $y$, so $x = 5$.

**25. e.** Let $x$ = the number of children and let $y$ = the number of adults. There are a total of 120 people, so $x + y = 120$. Since they need 20 gallons of water for each adult and 15 gallons for each child, then the total amount of water can be written as $15x + 20y = 1,950$. Use the elimination method to find $x$.

| | | |
|---|---|---|
| Multiply the first equation by $-20$: | $-20x + -20y$ | $= -2,400$ |
| Keep the second equation the same: | $\underline{15x + 20y}$ | $\underline{= 1,950}$ |
| Combine the two equations vertically to eliminate $y$: | $-5x$ | $= -450$ |
| Divide both sides of the equation by $-5$: | $\frac{-5x}{-5}$ | $= -\frac{-450}{-5}$ |

Therefore, $x = 90$. There are 90 children in the group.

# 5

# Linear and Compound Inequalities

**B**efore you begin learning and reviewing how to solve and graph linear and compound inequalities, take a few minutes to take this ten-question *Benchmark Quiz*. These questions are similar to the type of questions that you will find on important tests. When you are finished, check the answer key carefully to assess your results. Your Benchmark Quiz analysis will help you determine how much time you need to spend on this chapter and the specific areas in which you need the most careful review and practice.

## BENCHMARK QUIZ

1. Which of the following graphs represent the solution set of $x \leq 4$?

    **a.**

    **b.**

**c.**

```
←─┼──┼──┼──┼──●──┼──┼──→
  0  1  2  3  4  5  6
```

**d.**

```
←─┼──┼──┼──●──┼──┼──┼──→
  0  1  2  3  4  5  6
```

**e.**

```
←─┼──┼──┼──●──┼──┼──┼──→
  0  1  2  3  4  5  6
```

2. What inequality is represented by the graph?

```
←─┼──┼──┼──⊕──┼──┼──┼──┼──→
 -5 -4 -3 -2 -1  0  1
```

**a.** $y \leq -3$
**b.** $y \geq -3$
**c.** $y < -3$
**d.** $y > -3$
**e.** none of these

3. What is the solution of the inequality $3x > -9$
**a.** $x < 3$
**b.** $x > 3$
**c.** $x > -3$
**d.** $x < -3$
**e.** $x > \frac{1}{3}$

4. Which of the following is the solution of the inequality
$-6a - 4 \geq 8$?
**a.** $a \geq -2$
**b.** $a \leq -2$
**c.** $a \geq 2$
**d.** $a \leq 2$
**e.** $a < -2$

5. What compound inequality is shown in the graph?

```
←┼┼┼┼┼┼●┼┼┼┼┼┼┼┼┼⊕┼→
  -10    -5    0    5    10
```

**a.** $x \geq 9$
**b.** $-4 < x \leq 9$
**c.** $-4 \leq x < 9$
**d.** $x < -4$ or $x > 9$
**e.** $-4 < x < 9$

**6.** Which of the following is the graph of the compound inequality $x > 6$ or $x \leq -3$?

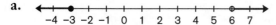

a.

b.

c.

d.

e.

**7.** What is the linear inequality represented by the graph?

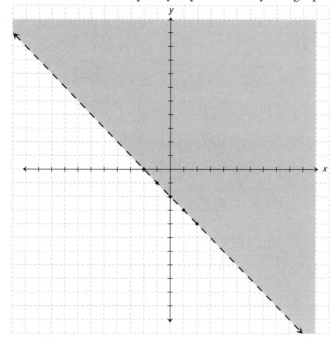

**a.** $y \leq x - 2$
**b.** $y < x - 2$
**c.** $y > -x - 2$
**d.** $y < -x - 2$
**e.** $y \geq -x - 2$

**8.** Which graph represents the inequality $5x \geq 4y - 8$?

**a.**

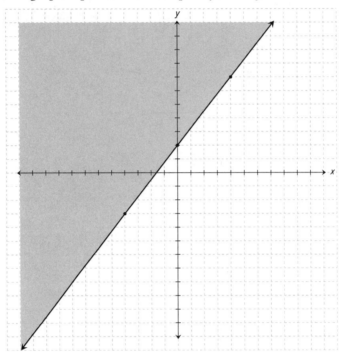

**b.**

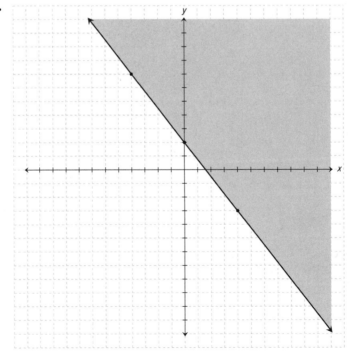

**c.**

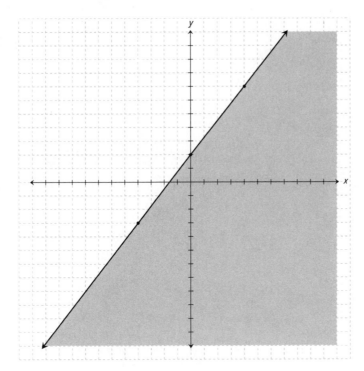

**d.**

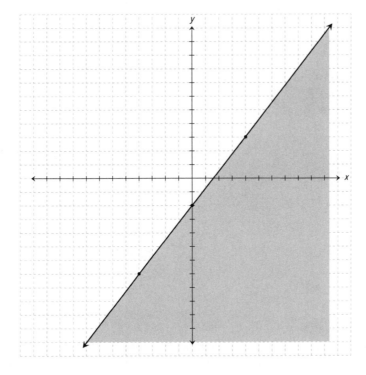

**e.**

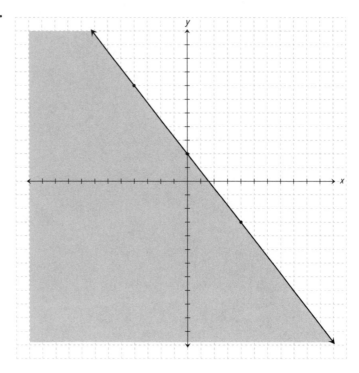

9. For which of the inequalities is the point (3,–2) a solution?
   **a.** $2y - x \geq 1$
   **b.** $x + y > 5$
   **c.** $3y < -3x$
   **d.** $9x - 1 > y$
   **e.** none of these

**10.** Which system of inequalities is represented in the graph?

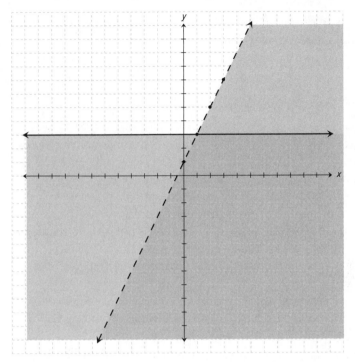

**a.** $y < 3$
   $y > 2x + 1$
**b.** $y < 2x + 1$
   $y \leq 3$
**c.** $y \leq 3$
   $y > -2x + 1$
**d.** $x < 3$
   $y > 2x + 1$
**e.** none of these

## BENCHMARK QUIZ SOLUTIONS

How did you do on solving and graphing inequalities? Check your answers here, and then analyze your results to figure out your plan to master these topics.

### ▶ Answers

**1. e.** You are looking for the solution to the inequality *x is less than or equal to 4*. This is a graph with a closed circle at 4 and the arrow pointing to the left. This figure is choice **e**.

**2. d.** In this graphic there is an open circle at –3 so the symbol in the inequality is either < or >. The arrow is pointing to the right toward numbers greater than –3. Therefore, the inequality representing the graph is *y is greater than –3*, which is written as $y > -3$.

**3. c.** Solve for *x* as you would in an equation. Divide each side of the inequality by 3; $\frac{3x}{3} > \frac{-9}{3}$. The solution is $x > -3$.

**4. b.** Solve for *a* as you would in an equation. Add 4 to both sides of the inequality; $-6a - 4 + 4 \geq 8 + 4$. Simplify; $-6a \geq 12$. Divide both sides of the inequality by –6; $\frac{-6a}{-6} \geq \frac{12}{-6}$. Remember to switch the direction of the inequality symbol because you are dividing both sides by a negative number; $a \leq -2$.

**5. c.** This graph shows that the solution set is all numbers between –4 and 9, and includes –4 because of the closed circle. In the solution set are numbers that are *greater than or equal to –4* and at the same time, *less than 9*. This is shown in the compound inequality in choice **c**.

**6. a.** The graph of a compound inequality shows two inequalities at once on the same graph. The inequality $x > 6$ has an open circle at 6 and arrow to the right. This only occurs in choices **a** and **d**. The other inequality, $x \leq -3$, has a closed circle at –3 and an arrow pointing to the left. The choice with both inequalities drawn correctly is choice **a**.

**7. c.** The line in the graph has a slope of –1 and a *y*-intercept of –2. In slope-intercept form this is $y = -x - 2$. Since the shading is above the line and the line itself is dashed, the equal sign should be replaced with a *greater than* sign. Therefore, the inequality for the graph is $y > -x - 2$, which is choice **c**.

**8. c.** Change the inequality to slope-intercept form. First, add 8 to both sides of the inequality; $5x + 8 \geq 4y - 8 + 8$. Simplify; $5x + 8 \geq 4y$. Then divide both sides of the inequality by 4; $\frac{5x}{4} + \frac{8}{4} \geq \frac{4y}{4}$. This simplifies to $\frac{5}{4}x + 2 \geq y$, which can also be written as $y \leq \frac{5}{4}x + 2$. The slope of the line is $\frac{5}{4}$, the $y$-intercept is 2, the line should be solid, and the shading is below the line. This occurs in choice **c**.

**9. d.** Substitute the point $(3, -2)$ into each answer choice to find the true inequality. Since the point is $(3, -2)$, the $x$-value is 3 and the $y$-value is $-2$. In choice **a** the inequality becomes $2(-2) - 3 \geq 1$. This simplifies to $-4 - 3 \geq 1$ which becomes $-7 \geq 1$. Negative seven is not greater than or equal to 1, so this choice is incorrect. In choice **b** the inequality becomes $3 + -2 > 5$ which simplifies to $1 > 5$. This is also not true so choice **b** is incorrect. In choice **c** the inequality becomes $3(-2) < -3(3,)$, which simplifies to $-6 < -9$. This is also not true. Choice **d** becomes $9(3) - 1 > -2$, which simplifies to $27 - 1 > 2$. Twenty-six is *greater than* 2, so choice **d** is the correct answer.

**10. b.** Take the graph one line at a time. The horizontal solid line at 3 is shaded below the line. This is the graph of the inequality $y \leq 3$. This inequality occurs in choices **b** and **c**. The diagonal line has a slope of 2 and a $y$-intercept of 1. This is an inequality in the form $y = 2x + 1$. Because the line is dashed and the shading is below the line, the inequality becomes $y < 2x + 1$. Both inequalities are in choice **b**.

## BENCHMARK QUIZ RESULTS

If you answered 8–10 questions correctly, you have a solid foundation in solving and graphing different types of inequalities. After reading through the lesson and focusing on the areas you need to refresh, try the quiz at the end of the chapter to ensure that all of the concepts are clear.

If you answered 4–7 questions correctly, you need to refresh yourself on some of the material. Read through the chapter carefully for review and skill building, and pay careful attention to the sidebars that refer you to more in-depth practice, hints, and shortcuts. Work through the quiz at the end of the chapter to check your progress.

If you answered 1–3 questions correctly, you need help and clarification on the topics in this section. First, carefully read this chapter and concentrate on these linear and compound inequality skills. Perhaps you learned this information and forgot it, so take the time now to refresh your skills and improve your knowledge. After taking the quiz at the end of the chapter, you may want to reference a more in-depth and comprehensive book, such as LearningExpress's *Algebra Success in 20 Minutes a Day*.

## JUST IN TIME LESSON—LINEAR AND COMPOUND INEQUALITIES

In this chapter you will be studying different types of inequalities and their respective solution sets. Specifically, you will review how to solve and graph:

- inequalities of one variable
- compound inequalities of one variable
- linear inequalities
- systems of linear inequalities

### ▶ *Solving Inequalities of One Variable*

 GLOSSARY

**INEQUALITY** a mathematical sentence that shows that two amounts are not equal

In many situations finding an exact quantity or amount is not necessary. You may be saving money to buy something and need *at least* $35. When you have saved $35 *or more* you have enough money, so there are many possible amounts that would satisfy this scenario. A situation such as this can be solved using inequalities, instead of equations. Some of the phrases you will see in this section are *at most*, *at least*, *less than*, and *greater than*, along with many others. This type of problem allows you to work with circumstances where finding just one solution is not enough.

The four symbols used when solving inequalities are:

< is less than
> is greater than
≤ is less than or equal to
≥ is greater than or equal to

Here are a few examples of simple inequalities:

4 > 3 is read "four is greater than three."
–2 ≤ –1 is read "negative two is less than or equal to negative one."

Solving an inequality is very similar to solving an equation. You will per-form the inverse operation of what you want to eliminate on both sides of the inequality. For example, in order to solve the inequality $5x \geq 25$, treat the *greater than or equal to* symbol like an equal sign. Divide both sides by 5 to get the $x$ alone; $\frac{5x}{5} \geq \frac{25}{5}$. This simplifies to $x \geq 5$. For the solution, $x$ is greater than or equal to 5. This means that $x$ could be any number 5 or greater.

The major difference between an equation and an inequality is the sym-bol used; most inequalities imply that more than one value of the variable will make the mathematical statement true. Therefore, when solving an inequality the solution set is graphed on a number line.

To graph a solution set, use the number in the solution as the starting point on the number line. In the problem $x \geq 5$, five is the starting point on the number line.

### ⌇⌇ RULE BOOK

**When using the symbol < or >, make an open circle at this number to show this number is not part of the solution set. When using the symbol ≤ or ≥, place a closed, or filled-in, circle at this number to show this num-ber is a part of the solution set.**

In the case of $x \geq 5$, the circle is closed. Next, draw an arrow from that point either to the left or the right on the number line. For the inequality $x \geq 5$, solutions to this problem are greater than or equal to five so the line should be drawn to the right.

Here are a few more examples of the graphs of solutions of different inequalities.

$x > 3$

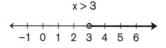

$x \leq 2$

$x < 4$

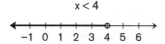

When solving inequalities, there is one catch:

For example, solve the inequality: $-3x + 6 \le 18$
First, subtract 6 from both sides: $-3x + 6 - 6 \le 18 - 6$
Then divide both sides by $-3$: $\frac{-3x}{-3} \le \frac{12}{-3}$
The inequality symbol now changes: $x \ge -4$

The graph of this solution is

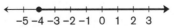

$$-5\ -4\ -3\ -2\ -1\ \ 0\ \ 1\ \ 2\ \ 3$$

*Be careful to watch for this type of situation.*

 RULE BOOK

**If you are multiplying or dividing each side of the inequality by a negative number, you must reverse the direction of the inequality symbol.**

To check a solution, choose a number in the solution set and substitute into the original inequality. To check the inequality $-3x + 6 \le 18$, check a value in the solution such as 0.

$$-3(0) + 6 \le 18$$
$$0 + 6 \le 18$$
$$6 \le 18$$

Since 6 is less than or equal to 18, this solution is checking.

## ▶ Word Problems

Here is an example of a word problem using inequalities.

Jim has twice as many pairs of socks as Debbie and together they have at least twelve pairs. What is the fewest number of pairs Debbie can have?

Let $x$ = the number of pairs Debbie has and let $2x$ = the number of pairs Jim has. Since the total pairs is *at least* 12, $x + 2x \ge 12$. Combine like terms on the left side; $3x \ge 12$. Divide both sides by 3; $\frac{3x}{3} \ge \frac{12}{3}$; $x \ge 4$. Debbie has at least 4 pairs of socks.

 SHORTCUT

**When solving problems with the words "at least," use the symbol $\ge$. When solving problems with the words "at most," use the symbol $\le$.**

## ▶ *Solving Compound Inequalities of One Variable*

A compound inequality is a combination of two or more inequalities. For example, take the compound inequality $-3 < x + 1 < 4$.

To solve this, subtract 1 from all parts of the inequality; $-3 - 1 < x + 1 - 1 < 4 - 1$. Simplify; $-4 < x < 3$.

Therefore, the solution set is all numbers between $-4$ and 3. On a graph the solution set is

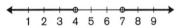

Here is another common type of a compound inequality:

Graph the solution set of $x < 4$ or $x > 7$.

SHORTCUT

The solution of a compound inequality will often be one of two general cases. One is between two values, such as the graph

These inequalities will sometimes use the word "and."

The other case is *less than or equal to* the smaller value or *greater than or equal to* the larger value as in the graph.

These inequalities will often use the word "or."

EXTRA HELP

For more information and practice solving inequalities of one variable, see *Algebra Success in 20 Minutes a Day*, Lesson 9: Solving Inequalities.

## ▶ *Graphing Linear Inequalities of Two Variables*

Graphing linear inequalities of two variables has basically the same steps as graphing linear equations. Put the inequality in slope-intercept form and identify the slope and the $y$-intercept. When you get ready to graph, however, there are two differences.

 RULE BOOK

 When graphing a linear inequality using the symbol < or >, the line drawn is dashed. Points on the line *are not* part of the solution set. Inequalities with ≤ or ≥ are solid lines. Points on those lines *are* part of the solution set.

   The other difference is that one side of the line is shaded to show all points in the plane that satisfy the inequality. Keep in mind that most inequalities have many, even infinite solutions. Shading one side of a linear inequality is called shading the *half-plane*.
   To find the side of the line that should be shaded, use a test point. If the test point makes the inequality true, then shade the side of the line where the test point is located. If the test point makes the inequality false, shade the opposite side of the line.

SHORTCUT

A good test point to use is (0,0) because it is easy to substitute in and evaluate. The only time a test point of (0,0) should not be used is if it is contained exactly on the line. In that case you should choose a different test point.

Here is an example of graphing a linear inequality.

   Graph the inequality $y < -2x + 1$.
   This inequality is in slope-intercept form; the slope is $-2$ and the $y$-intercept is 1. Before you begin graphing, look at the inequality symbol. Since the symbol is less than, the line should be a dashed line. Draw the line on an $x$–$y$ axis as you would a linear equation.

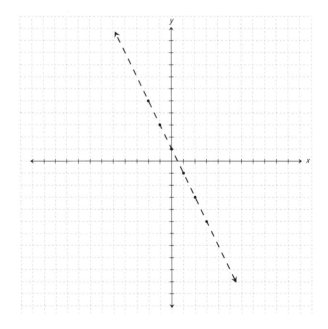

Now, use a test point to determine which side of the line should be shaded. Since (0,0) is not contained on the line, use that point; $y <$ $-2x + 1$ becomes $0 < -2(0) + 1$. This simplifies to $0 < 0 + 1$ which is equal to $0 < 1$. Since this is a true statement, the point (0,0) is in the solution set of the inequality. On the graph, shade the side of the line that contains the origin. This is the side below the line. *All* points on this side of the inequality are solutions.

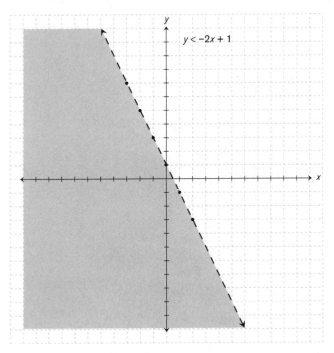

$y < -2x + 1$

## SHORTCUT

**In general, for inequalities using $<$ and $\leq$ shade below the line. For inequalities using $>$ and $\geq$ shade above the line.**

## ▶ *Special Cases of Linear Inequalities*

Two special cases of linear inequalities are horizontal lines and vertical lines.

**Horizontal lines** have a slope of zero and are in the form $y < k$ or $y > k$, where $k$ is any number.

The graph of the inequality $y < 2$ is

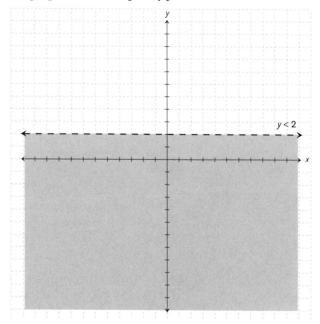

The graph of the inequality $y > -6$ is

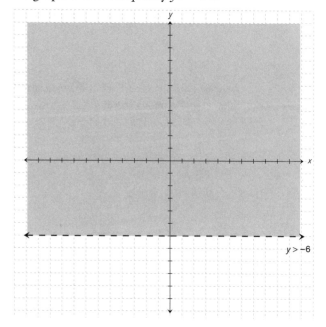

**Vertical lines** have undefined slope and are in the form $x < k$ or $x > k$, where $k$ is any number.

The graph of the inequality $x < 3$ is

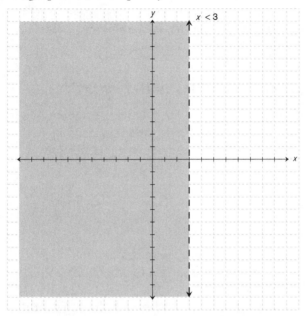

The graph of the inequality $x > -2$ is

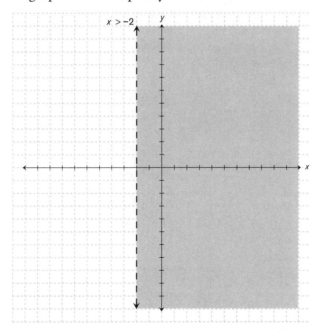

## EXTRA HELP

Check out the website www.math.com for computer-aided practice on graphing inequalities and systems. Under **Select Tool** on the home page, click on **Graphers**, then **Graph Inequalities**. Select **Plot Inequalities** for instant reinforcement of graphing skills. For additional information and practice graphing inequalities, see *Algebra Success in 20 Minutes a Day*, Lesson 10: Graphing Inequalities.

## ▶ Solving Systems of Inequalities in Two Variables

Solving a system of inequalities is similar to solving a system of equations. The difference here is that you are looking for an overlapping in the shaded portions of the inequalities, not just one point of intersection.

*Example:*
Graph the solution of the system of linear inequalities.

$$y + x > 3$$
$$2y \leq 4x + 2$$

In order to graph the solution to the system, graph both inequalities on the same set of axes and look for the region on the graph where the shaded sections overlap.

In the first inequality, subtract $x$ from both sides to get the inequality in slope-intercept form. This inequality becomes $y > -x + 3$ with a slope of $-1$ and a $y$-intercept of 3. Graph this line on a set of axes making the line dashed. Shade above the line because of the $>$ symbol.

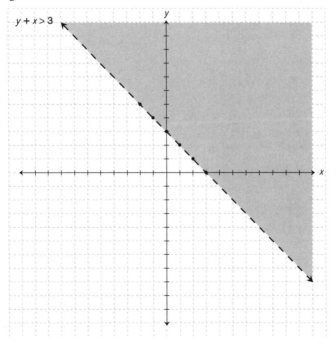

In the second inequality divide both sides by 2 to get it into slope-intercept form. The result is the inequality $y \leq 2x + 1$ with a slope of 2 and a $y$-intercept of 1. Draw a solid line and shade the region below the line.

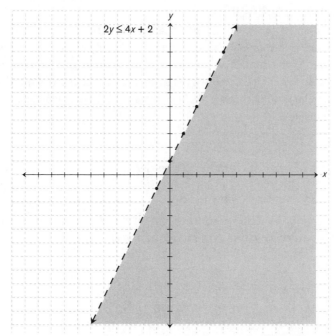

$2y \leq 4x + 2$

Now graph both lines on the same set of axes.

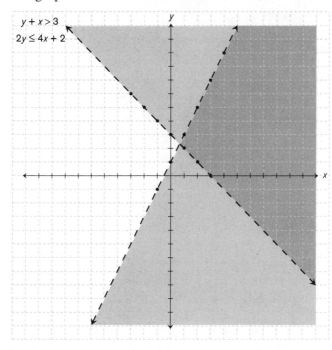

$y + x > 3$

$2y \leq 4x + 2$

The region where the shaded areas overlap is labeled **S** in the figure. Any ordered pair in this region is a solution to the system of inequalities.

## CALCULATOR TIP

Inequalities can be graphed on a graphing calculator using the **Y =** screen. Type your inequality in as you would an equation and move your cursor to the far left of the equal sign. Use the **ENTER** key to toggle through the choices. Choose the symbol for shading up for *greater than* inequalities and choose the symbol for shading down for *less than* inequalities.

A special type of system of inequalities occurs if the lines are parallel. Take, for instance, the system of inequalities $y > x + 3$ and $y < x - 1$. For the first inequality, graph a dashed line with slope of 1 and $y$-intercept of 3 and shade above the line. For the second inequality, graph a dashed line with slope of 1 and $y$-intercept of $-1$ and shade below the line.

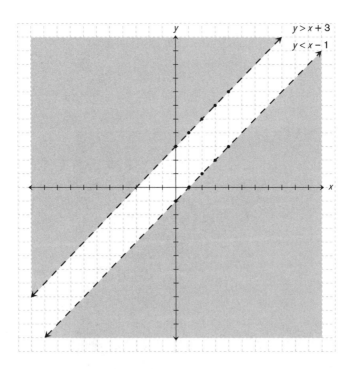

Notice that on the graph there is not an intersection of the shaded areas. Therefore, there is no solution to this system of inequalities.

## EXTRA HELP

For additional information and practice solving systems of inequalities, see *Algebra Success in 20 Minutes a Day*, Lesson 12: Graphing Systems of Linear Equations and Inequalities.

## TIPS AND STRATEGIES

Solving and graphing different types of inequalities doesn't have to be difficult. Here are a couple of things to keep in mind as you are improving your skills.

- Most of the steps to solving inequalities are the same steps you use to solve equations.
- Remember to change the direction of the inequality symbol when dividing or multiplying each side of the inequality by a negative value.
- When graphing the solution set of a one variable inequality, use an open circle in problems with the symbols < and >. Use a closed circle in problems with the symbols ≤ and ≥.
- When graphing any solution set, use a test point to ensure that the arrow points in the right direction or that the correct half-plane is shaded.
- A compound inequality usually involves two inequalities at one time. Solve them one at a time to make it easier.
- Linear inequalities using the symbols < or > should be dashed lines when graphed.
- In general, shade above the line for a *greater than* inequality.
- In general, shade below the line for a *less than* inequality.
- To solve a system of linear inequalities, graph both inequalities on the same set of axes and look for the region where the shaded areas overlap.
- Use process of elimination to answer inequality multiple-choice questions. Look for traits such as an open circles, or a dashed line on a graph to match a given inequality.

## CHAPTER QUIZ

Try these practice problems to track your progress through solving and graphing inequalities.

**1.** Which of the following graphs represent the solution set of $x > -1$?

**a.**
-3 -2 -1  0  1  2

**b.**
-3 -2 -1  0  1  2

**c.**
-3 -2 -1  0  1  2

**d.**
-3 -2 -1  0  1  2

**e.**
-3 -2 -1  0  1  2

**2.** What inequality is represented by the graph?

3  4  5  6  7  8  9

**a.** $x \leq 6$
**b.** $x \geq 6$
**c.** $x < 6$
**d.** $x > 6$
**e.** none of these

**3.** What is the solution set of the inequality $x - 3 > -8$?
**a.** $x < -5$
**b.** $x > -5$
**c.** $x < -11$
**d.** $x > -11$
**e.** $x > 11$

**4.** Which of the following is the solution of the inequality
$-\frac{a}{2} - 10 \geq -2$?
**a.** $a \geq -16$
**b.** $a \leq 1$
**c.** $a \leq -1$
**d.** $a \geq 1$
**e.** $a \leq -16$

5. Which of the following is NOT in the solution set of $x - 1 \le -9$?
   a. $-8$
   b. $-9$
   c. $-7$
   d. $-21$
   e. $-10$

6. Solve the inequality: $x + 7 > 11$.
   a. $x > 4$
   b. $x < 4$
   c. $x < 18$
   d. $x > 18$
   e. $x > \frac{11}{7}$

7. Solve the inequality: $9 - 5x \ge 39$.
   a. $x > 6$
   b. $x \ge -6$
   c. $x > -6$
   d. $x \ge 6$
   e. $x \le -6$

8. Which of the following is the solution to the inequality
   $5 + 7(3 - 5x) \le 96$?
   a. $x < -2$
   b. $x \le 2$
   c. $x \le -2$
   d. $x \ge 2$
   e. $x \ge -2$

9. Charles has three more than twice as much money as Betsy has. What
   is the maximum amount of money Betsy has if together they have at
   most $48?
   a. 15
   b. 17
   c. 33
   d. 37
   e. 47

**10.** Six more than twice a number is at least 38. What is the smallest value of the number?
   **a.** 10
   **b.** 14
   **c.** 16
   **d.** 22
   **e.** 32

**11.** What compound inequality is shown in the graph?

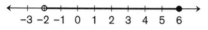

-3 -2 -1  0  1  2  3  4  5  6

                                 **a.**        $x \geq 6$
   **b.** $-2 < x \leq 6$
   **c.** $-2 \leq x < 6$
   **d.** $x < -2$
   **e.** $-2 < x < 6$

**12.** Which of the following is the graph of the compound inequality $x \leq 6$ or $x > 8$?

   **a.**
   3  4  5  6  7  8  9

   **b.**
   3  4  5  6  7  8  9

   **c.**
   3  4  5  6  7  8  9

   **d.**
   3  4  5  6  7  8  9

   **e.**
   3  4  5  6  7  8  9

13. Solve the compound inequality: $3 < x - 2 < 4$.
    a. $1 < x < 2$
    b. $5 < x < 6$
    c. $1 > x > 5$
    d. $5 > x > 6$
    e. $3 < x < 4$

14. The following graph is the solution set to which inequality?

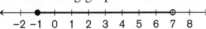

    a. $-1 \leq a > 7$
    b. $-1 < a < 7$
    c. $-1 \leq a < 7$
    d. $-1 \geq a > 7$
    e. $-1 < a > 7$

15. Match the graph with the inequality: $x > 3$.
    a.

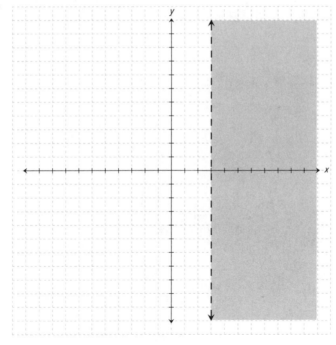

**b.**

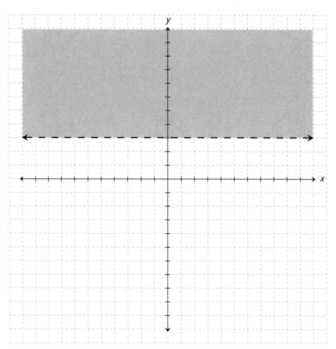

**c.**

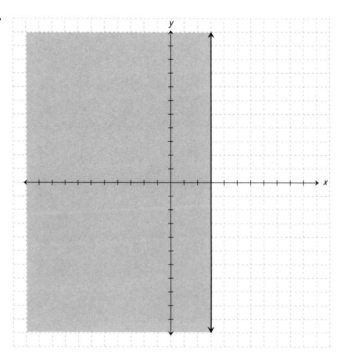

**d.**

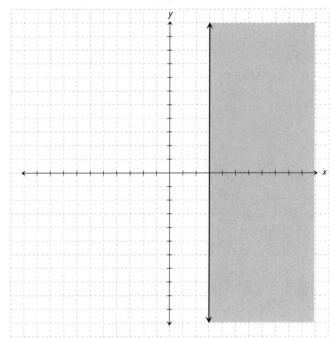

**e.**

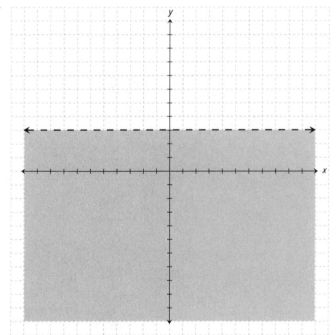

**16.** Match the graph with the inequality: $y \le 4x - 1$.

**a.**

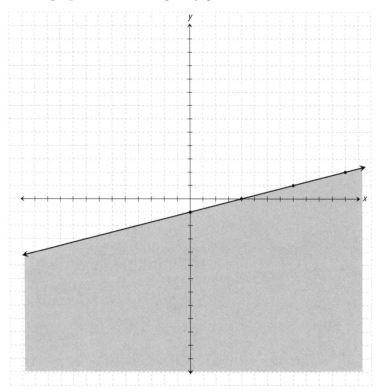

**b.**

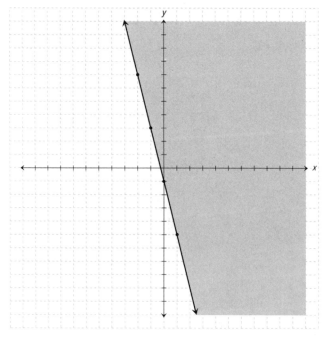

**c.**

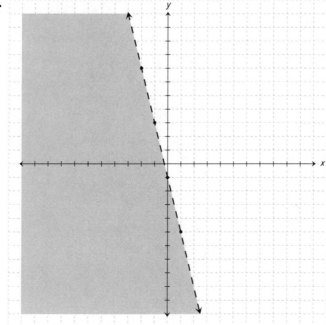

**d.**

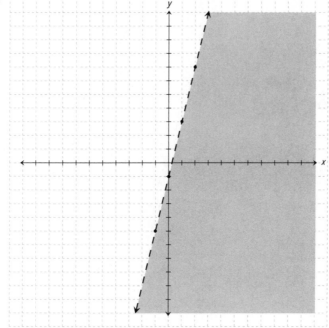

**e.**

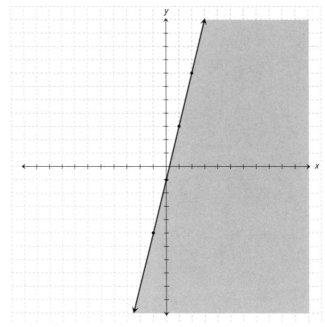

**17.** What is the linear inequality represented by the graph?

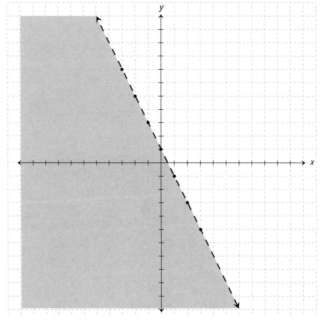

**a.** $y \le -2x + 1$
**b.** $y < -2x + 1$
**c.** $y > -2x + 1$
**d.** $y \le -2x + 1$
**e.** $y \ge -2x + 1$

**18.** Which ordered pair is in the solution set of the inequality
$3x \geq -y - 6$?
  **a.** $(10,-10)$
  **b.** $(1,-10)$
  **c.** $(-3,-2)$
  **d.** $(-5,1)$
  **e.** $(0,-7)$

**19.** For which of the inequalities is the point $(-4,-5)$ a solution?
  **a.** $2y - 2x \geq 1$
  **b.** $x + y > 3$
  **c.** $y < -3x$
  **d.** $5x - 2 > y$
  **e.** none of these

**20.** Match the inequality with the graph.

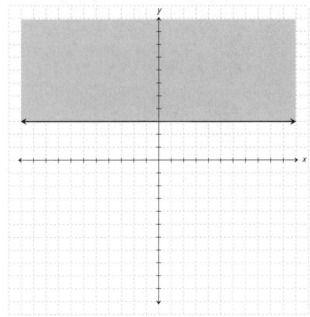

  **a.** $y \leq 3$
  **b.** $y > 3$
  **c.** $x < 3$
  **d.** $y \geq 3$
  **e.** $x \leq 3$

**21.** For which inequality is the ordered pair (10,–2) NOT a solution?

    **a.** $x \geq 2y$

    **b.** $-y \geq x - 8$

    **c.** $y > x - 12$

    **d.** $3x > 3y + 1$

    **e.** $x > y$

**22.** Which system of inequalities is represented in the graph?

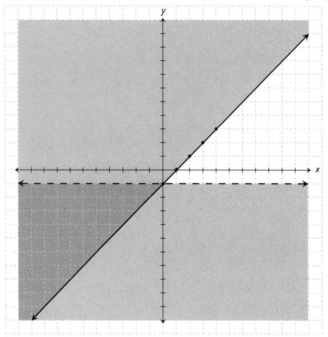

    **a.** $y > -1$

       $y \geq x - 1$

    **b.** $y \leq -1$

       $y \geq x - 1$

    **c.** $y < -1$

       $y \leq x - 1$

    **d.** $y \geq -1$

       $y < x - 1$

    **e.** none of these

23. For which system of inequalities is the point (0,0) in the solution set?
   a. $y > 3$
   $x > 2$
   b. $y \geq 4$
   $x > 1$
   c. $y \geq -3$
   $x < 1$
   d. $y < -1$
   $x < -2$
   e. none of these

24. For which of the following systems is there no solution?
   a. $y < -1$
   $x < 2$
   b. $y < 3x$
   $y > -3x$
   c. $y \geq -x + 1$
   $y \leq x + 1$
   d. $y \leq -2$
   $y \geq 2$
   e. $y \geq -2x$
   $x > 1$

**25.** Match the graph with the system of inequalities.

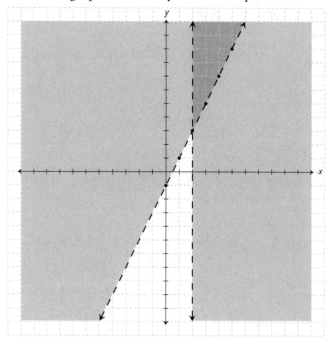

**a.** $y > 2x - 1$
   $x > 2$
**b.** $y > 2x - 1$
   $x < 2x$
**c.** $y \le -2x - 1$
   $x < 2$
**d.** $y < 2x - 1$
   $x > 2$
**e.** none of these

## ANSWERS

Following are the answers and explanations for the chapter quiz. Read over the explanations carefully to correct any misunderstandings. Refer to LearningExpress's *Algebra Success in 20 Minutes A Day* for further review and practice.

**1. b.** The inequality $x > -1$ is read "*x* is *greater than* $-1$." On a graph, this means an open circle at $-1$ and an arrow pointing to the right. This is choice **b**.

**2. a.** In the graph there is a closed circle at 6 and an arrow pointing to the left, which are values less than 6. This is the inequality $x \leq 6$, which is choice **a**.

**3. b.** Solve the inequality $x - 3 > -8$ as you would an equation. Add 3 to both sides of the inequality; $x - 3 + 3 > -8 + 3$. This simplifies to $x > -5$, which is choice **b**.

**4. e.** Solve the inequality as you would an equation. Add 10 to both sides of the inequality; $-\frac{a}{2} - 10 + 10 \geq -2 + 10$. Simplify; $-\frac{a}{2} \geq 8$. Multiply each side by $-2$; $\frac{-a}{2} \bullet -2 \geq 8 \bullet -2$. Simplify and switch the direction of the inequality symbol; $a \leq -16$.

**5. c.** Solve the inequality for *x*. Add 1 to both sides of the inequality. $x - 1 + 1 \leq -9 + 1$. This simplifies to $x \leq -8$. Any number less than or equal to $-8$ is a solution to the inequality. Therefore $-7$ is not in the solution set.

**6. a.** Solve the inequality for *x*. Subtract 7 from each side of the inequality; $x + 7 - 7 > 11 - 7$. This simplifies to $x > 4$.

**7. e.** Solve the inequality for *x*. Subtract 9 from both sides of the inequality; $9 - 9 - 5x \geq 39 - 9$. Simplify; $-5x \geq 30$. Divide both sides of the inequality by $-5$. Remember that when dividing or multiplying each side of an inequality by a negative number, the inequality symbol changes direction; $\frac{-5x}{-5} \leq \frac{30}{-5}$. The variable is now alone; $x \leq -6$.

**8. e.** Change subtraction to addition and the sign of the number following to its opposite; $5 + 7(3 + -5x) \leq 96$. Use the distributive property on the left side of the inequality; $5 + 21 + -35x \leq 96$.

Combine like terms on the left side of the inequality; $26 + -35x \le 96$. Subtract 26 from both sides of the inequality; $26 - 26 + -35x \le 96 - 26$. Simplify; $-35x \le 70$. Divide both sides of the inequality by $-35$. Don't forget that when you divide or multiply each side of an inequality by a negative number, the direction of the inequality changes; $\frac{-35x}{-35} \ge \frac{70}{-35}$; $x \ge -2$.

**9. a.** Let $x$ = the amount of money Betsy has. Let $2x + 3$ = the amount of money Charles has. Since the total money added together was *at most* 48, the inequality would be $(x) + (2x + 3) \le 48$. Combine like terms on the left side of the inequality; $3x + 3 \le 48$. Subtract 3 from both sides of the inequality; $3x + 3 - 3 \le 48 - 3$. Simplify; $3x \le 45$. Divide both sides of the inequality by 3; $\frac{3x}{3} \le \frac{45}{3}$. The variable is now alone. $x \le 15$. The maximum amount of money Betsy has is $15.

**10. c.** Let $x$ = a number. Then six more than twice a number is equal to $2x + 6$. Since this amount is *at least* 38, then the inequality is $2x + 6 \ge 38$. Subtract 6 from both sides of the inequality; $2x + 6 - 6 \ge 38 - 6$. Simplify; $2x \ge 32$. Divide each side by 2; $\frac{2x}{2} \ge \frac{32}{2}$; $x \ge 16$. The minimum value for $x$ is 16.

**11. b.** The graph has an open circle at $-2$, a closed circle at 6, and shows values between $-2$ and 6. So $x$ is *greater than* $-2$ $(x > -2)$ and *less than or equal to* 6 $(x \le 6)$ at the same time. The combination of these two inequalities is in choice **b**.

**12. d.** The graph of $x \le 6$ has a closed circle at 6 and arrow to the left. The graph of $x > 8$ has an open circle at 8 and arrow to the right. Both inequalities are graphed correctly in choice **d**.

**13. b.** This problem is an example of a compound inequality, where there is more than one inequality in the question. In order to solve it treat it as two separate inequalities; $3 < x - 2 < 4$ becomes $3 < x - 2$ and $x - 2 < 4$. Add 2 to both sides of both inequalities; $3 + 2 < x - 2 + 2$ and $x - 2 + 2 < 4 + 2$. Simplify; $5 < x$ and $x < 6$. If $x$ is greater than five and less than six, it means that the solution is between 5 and 6 and can be shortened to $5 < x < 6$.

**14. c.** The graph of the inequality shows a solution set where $a$ is *greater than or equal to* $-1$ by the closed circle at $-1$ and also *less than* 7. The graph does not include 7 because of the open circle at 7. The graph connects the two circles; therefore the numbers between $-1$

and 7 are in the solution set; $a \geq -1$ and $a < 7$ can also be written as $-1 \leq a < 7$.

**15. a.** The line $x > 3$ is a vertical dashed line where $x = 3$. This is choice **a**.

**16. e.** The line $y \leq 4x - 1$ is a solid line with a slope of 4 and a $y$-intercept of $-1$. Because the symbol in the inequality is *less than or equal to*, the shading should be below the line. This is correctly drawn in choice **e**.

**17. b.** The line drawn in the graph has a slope of $-2$ and a $y$-intercept of 1. The line is dashed, so the symbol for the inequality is either $<$ or $>$. Since the area below the line is shaded, the correct inequality is $y < -2x + 1$.

**18. a.** One way to solve this problem is to substitute each ordered pair into $3x \geq -y - 6$ and find the true inequality. In choice **a**, the inequality becomes $3(10) \geq - (-10) - 6$. This simplifies to $30 \geq 10 - 6$ which is equal to $30 \geq 4$. This is true, so choice **a** is the answer. Each of the other inequalities is false. Choice **b** becomes $3(1) \geq - (-10) - 6$ which is equal to $3 \geq 4$. Choice **c** becomes $3(-3) \geq - (-2) - 6$ which is equal to $-9 \geq -4$. Choice **d** becomes $3(-5) \geq -1 - 6$ which is equal to $-15 \geq -7$. Choice **e** becomes $3(0) \geq - (-7) - 6$ which is equal to $0 \geq 1$. Another way to solve this problem is to graph the original inequality and the points from each of the answer choices. Only the point $(10, -10)$ would be located in the shaded region of the inequality.

**19. c.** Substitute the point $(-4, -5)$ into each inequality and look for the true statement. For choice **a**, the inequality becomes $2(-5) - 2(-4) \geq 1$. This simplifies to $-10 - (-8) \geq 1$ which then becomes $-2 \geq 1$. This is not true. Choice **b** becomes $-4 + -5 > 3$ which is equal to $-9 > 3$. This is also not true. Choice **c** becomes $-5 < -3(-4)$ which simplifies to $-5 < 12$. This is a true statement and is the solution to the problem. Choice **d** is not true. Substitute to get $5(-4) - 2 > -5$ which becomes $-20 - 2 > -5$; $-22$ is not greater than $-5$.

**20. d.** The line drawn is a horizontal line, so the form of the inequality will use $y$ only. The area above the line is shaded, so you are looking for a *greater than* or *greater than or equal to* symbol, as in choices **b** and **d**. Since the line is solid it is *greater than or equal to* as in choice **d**.

**21. c.** In this case, you are looking for the inequality that is NOT true. Substitute the point (10,–2) into each answer choice. Choice **a** becomes $10 \geq 2(-2)$ which is equal to $10 \geq -4$. This statement is true. Choice **b** becomes $-(-2) \geq 10 - 8$ which is equal to $2 \geq 2$. This is also true. Choice **c** becomes $-2 > 10 - 12$, which is equal to $-2 > -2$. This is not true, and is the solution to the problem. Choice **d** becomes $3(10) > 3(-2) + 1$ which is equal to $30 > -5$. This is true. Choice **e** becomes $10 > -2$, which is also true.

**22. e.** The dashed horizontal line is at $y = -1$ and is shaded below the line. Therefore the inequality is $y < -1$. The solid diagonal line has slope of 1 and a $y$-intercept of $-1$. It is shaded above the line. The inequality for this line is $y \geq x - 1$. These two inequalities are not listed together as an answer choice, so the correct answer is **e**.

**23. c.** Take each inequality and substitute $x = 0$ and $y = 0$. In choices **a, b,** and **d,** both inequalities are false. In choice **c,** both inequalities are true; $0 \geq -3$ and $0 < 1$. Another way to solve this problem is to graph each system. The only system with the origin in the solution set is choice **c.**

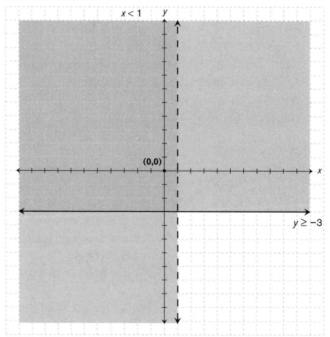

**24. d.** To find a system of inequalities with no solution, first look for two parallel lines. Recall that any two lines will different slope will eventually intersect. Then check to see that the shaded regions of the lines do not overlap. Choice **a** has a vertical and a horizontal line. There will be a solution to this system. Choices **b** and **c** both have two inequalities with different slopes. Each will have a solution. Choice **d** is a system of two horizontal lines, so they are parallel. Because of the inequality symbols used in this problem, the graph of the system will have not have any overlapping of shaded regions.

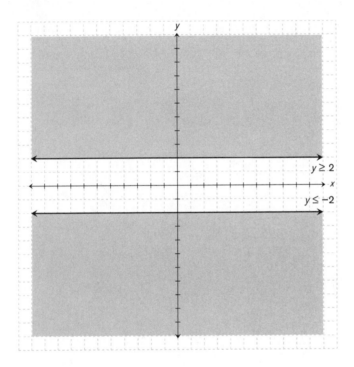

There is no solution to this system. Choice **e** contains a diagonal line and a vertical line. There will be a solution to the system stated in choice **e**.

**25. a.** Both of the lines in the graph are dashed. The diagonal line in the graph has slope of 2 and a $y$-intercept of $-1$. The shading is above the line. The inequality that matches this line is $y > 2x - 1$. The vertical line in the graph is located at $x = 2$ and the shading is to the right. The inequality that matches this graph is $x > 2$. The answer choice with both inequalities stated is choice **a.**

**6**

# Exponents

**B**efore beginning your study of exponents, take a few minutes to take this ten-question *Benchmark Quiz*. The types of questions you will encounter on the quiz are similar to the types you will see on important tests. There is an answer key to check your answers at the end of the quiz. Use the explanations to identify errors and to carefully assess your results. Your Benchmark Quiz analysis will help you determine how much time you need to spend on operations with exponents and the specific areas in which you need practice and review the most.

## BENCHMARK QUIZ

1. Evaluate $6^4$.
   a. 10
   b. 24
   c. 36
   d. 216
   e. 1,296

2. Which of the following is NOT equivalent to $5 \bullet 5 \bullet 5 \bullet 5$?
   a. $5 \bullet 4$
   b. $(5 \bullet 5)^2$
   c. $5^4$
   d. $5 \bullet 5^3$
   e. 625

3. Evaluate $cd^2 - 1$ when $c = -1$ and $d = -6$.
   a. $-36$
   b. 35
   c. $-37$
   d. $-13$
   e. 24

4. The expression $4^{-2}$ is equivalent to
   a. $4^2$
   b. $\frac{1}{4^{-2}}$
   c. 16
   d. $\frac{1}{16}$
   e. $-16$

5. Which of the following is equivalent to $2^2 \bullet 2^3$?
   a. 2
   b. $2^5$
   c. $2^6$
   d. $4^5$
   e. $4^6$

6. Simplify: $a^2b \bullet ab^3$.
   a. $a^2b^3$
   b. $ab^2$
   c. $a^3b^4$
   d. $2ab^3$
   e. $ab^5$

7. Simplify: $\frac{x^6}{x^3}$.
   a. $x^2$
   b. $x^3$
   c. $x^{-2}$
   d. $x^9$
   e. $x^{18}$

**8.** Simplify $(3xy^3)^2$.
   **a.** $3xy^5$
   **b.** $3xy^6$
   **c.** $3x^2y^6$
   **d.** $9x^2y^6$
   **e.** $9x^3y^5$

**9.** Which of the following is equivalent to $\sqrt{128}$?
   **a.** $12\sqrt{8}$
   **b.** $2\sqrt{8}$
   **c.** $8\sqrt{2}$
   **d.** $16\sqrt{8}$
   **d.** $64\sqrt{2}$

**10.** Which of the following is equivalent to $-3\sqrt{2} \bullet 7\sqrt{5}$?
   **a.** $-10\sqrt{7}$
   **b.** $-10\sqrt{10}$
   **c.** $-21\sqrt{7}$
   **d.** $-21\sqrt{10}$
   **e.** $-\sqrt{210}$

## BENCHMARK QUIZ SOLUTIONS

What did you remember about exponents and radicals? Check your answers here, and then analyze your results to plan your study and review of these topics.

### ▶ Answers

**1. e.** $6^4$ is equal to $6 \bullet 6 \bullet 6 \bullet 6$. When multiplied together the result is 1,296.

**2. a.** Each of the following choices are equivalent except $5 \bullet 4$. This is equal to 20. The remainder of the answer choices are equivalent to 625.

**3. c.** Substitute the values for the variables in the expression; $(-1)(-6)^2 - 1$. Evaluate the exponent; $(-1)(-6)^2 - 1$. Remember that $(-6)^2 = (-6)(-6) = 36$. Multiply the first term; $(-1)(36) - 1$. This simplifies to $(-36) - 1$. Evaluate by changing subtraction to addition and the sign of the second term to its opposite. Signs are the same so add and keep the sign; $(-36) + (-1) = -37$.

**4. d.** When evaluating a negative exponent, take the reciprocal of the base and make the exponent positive. Therefore, $4^{-2}$ is equivalent to $\frac{1}{4^2}$ which simplifies to $\frac{1}{16}$.

**5. b.** When multiplying like bases, add the exponents. The expression $2^2 \cdot 2^3$ is equivalent to $2^{2+3}$ which simplifies to $2^5$.

**6. c.** When multiplying like bases, add the exponents. The expression $a^2b \cdot ab^3$ can also be written as $a^2b^1 \cdot a^1b^3$. Grouping like bases results in $a^2a^1 \cdot b^1b^3$. Adding the exponents gives $a^{2+1}b^{1+3}$ which is equal to $a^3b^4$, the simplified answer.

**7. b.** When dividing like bases, subtract the exponents. The expression $\frac{x^6}{x^3}$ then becomes $x^{6-3}$ which simplifies to $x^3$.

**8. d.** When raising a quantity to a power, raise each base to that power by multiplying the exponents. The expression $(3xy^3)^2$ equals $3^2x^2y^6$ which simplifies to $9x^2y^6$. Another way to look at this problem is to remember that when a quantity is squared, it is multiplied by itself. The expression $(3xy^3)^2$ becomes $(3xy^3) \cdot (3xy^3)$. Multiply coefficients and add the exponents of like bases; $3 \cdot 3x^{1+1}y^{3+3}$ simplifies to $9x^2y^6$.

**9. c.** To simplify a radical, find the largest perfect square factor of the radicand. Since 128 can be expressed as $64 \cdot 2$, write $\sqrt{128}$ as $\sqrt{64} \cdot \sqrt{2}$. Since $\sqrt{64}$ equals 8, the radical reduces to $8\sqrt{2}$.

**10. d.** When multiplying radicals, multiply the numbers in front of the radicals together and then the radicands together. For the expression $-3\sqrt{2} \cdot 7\sqrt{5}$, multiply $-3 \cdot 7$ and $\sqrt{2} \cdot \sqrt{5}$. This simplifies to $-21\sqrt{10}$.

## BENCHMARK QUIZ RESULTS

If you answered 8–10 questions correctly, you have a solid foundation of the principles of working with exponents and radicals. After reading through the lesson and focusing on the areas you need to review, try the quiz at the end of the chapter to ensure that all of the concepts are clear.

If you answered 4–7 questions correctly, you need to refresh yourself on some of the material. Read through the chapter carefully for review and skill building, and pay careful attention to the sidebars that refer you to more in-depth practice, hints, and shortcuts. Work through the quiz at the end of the chapter to check your progress.

If you answered 1–3 questions correctly, there are a number of skills in this unit you should review. Take the time to carefully read through the chapter, highlighting areas of misunderstanding. Use the sidebars for extra clarification of skills and definitions. When your study of the chapter is complete, work through the quiz at the end of the chapter. You may be surprised at the improvement practice and extra explanations can bring. In addition to the quiz, you may want to reference a more in-depth and comprehensive book, such as LearningExpress's *Algebra Success in 20 Minutes a Day*.

## JUST IN TIME LESSON—EXPONENTS

This lesson will cover important properties and applications of exponents and radicals. Specifically you will review:

- how to simplify and evaluate various types of exponents
- how to simplify radicals and radical expressions
- how to perform operations with radicals

### ▶ Exponents

The exponent of a number tells how many times to use that number as a factor. For example, in the expression $4^3$, four is the *base number* and three is the *exponent*, or *power*.

$$4^3 \leftarrow \text{exponent}$$
$$\nwarrow \text{base}$$

Four should be used as a factor three times: $4^3 = 4 \bullet 4 \bullet 4 = 64$.
In a similar way, the expression $5^2$ is equal to $5 \bullet 5$ which is 25.

 RULE BOOK

**If there is no exponent written with a base number or variable, the exponent is assumed to be 1. For example, $5 = 5^1$ and $x = x^1$.**

When evaluating the order of operations with exponents, working out the powers comes very early in the problem, just after parentheses. For example when evaluating the expression $x^2y - 5$ for $x = 2$ and $y = -5$, first substitute the values for $x$ and $y$. The expression becomes $(2)^2 \bullet (-5) - 5$. Since there are no values inside the parentheses to be simplified, evaluate the exponent on the base of 2. The expression is now $4 \bullet -5 - 5$. Multiply to get a result of $-20 - 5$. Subtract for a final answer of $-25$.

## CALCULATOR TIP

When using a scientific or graphing calculator, there are multiple ways to evaluate exponents. To square a number, or raise it to a power of 2, use the $x^2$ button. To evaluate exponents other than 2, use the power key which looks like

$$\boxed{y^x} \quad \text{or} \quad \boxed{x^y}$$

For example, to calculate the value of $5^3$ press the key sequence

$$\boxed{5} \;\; \boxed{y^x} \;\; \boxed{3}$$

The answer is 125. If your particular calculator has the button

$$\boxed{\wedge}$$

to raise a number to an exponent, use it in the same manner. For example, $3^4$ would be entered

$$\boxed{3} \;\; \boxed{\wedge} \;\; \boxed{4}$$

The result is 81.

In the same fashion, exponents can be used in algebraic expressions. In the expression $x^4$, $x$ is the base and 4 is the exponent. In expanded form this means $x \bullet x \bullet x \bullet x$. The expression $ab^2$ represents $a \bullet b \bullet b$ in expanded form. The variable $a$ is a base with an exponent of 1 and $b$ is a base with an exponent of 2.

## RULE BOOK

Any non-zero number with zero as the exponent is equal to one. For example, $14^0 = 1$.

## ▶ Laws of Exponents

There are five rules that can be used when performing operations with exponents.

1. **Multiplying Like Bases.** The first rule mentioned here deals with multiplying *like* base numbers as in $3^4 \bullet 3^2$. If $3^4 \bullet 3^2$ is written in expanded form it is equal to $(3 \bullet 3 \bullet 3 \bullet 3) \bullet (3 \bullet 3)$. This quantity can also be written as $3^6$. Three is used as a factor six times using the exponents, $4 + 2 = 6$. Therefore, when the bases are the same, add the exponents to simplify.

RULE BOOK

**When multiplying like bases, add the exponents:** $x^a \cdot x^b = x^{a+b}$.

2. **Dividing Like Bases.** When dividing in an expression such as $\frac{x^5}{x^2}$, first use expanded form to simplify;

$$\frac{x^5}{x^2} = \frac{x \cdot x \cdot x \cdot x \cdot x}{x \cdot x} = \frac{\overset{1}{\cancel{x}} \cdot \overset{1}{\cancel{x}} \cdot x \cdot x \cdot x}{\underset{1}{\cancel{x}} \cdot \underset{1}{\cancel{x}}} = x^3.$$

Notice that in a division expression using *like* bases with exponents of 5 and 2, the result is an exponent of 3. When performing division with like bases you should subtract the exponents; $\frac{x^5}{x^2} = x^{5-2} = x^3$.

 RULE BOOK

**When dividing like bases, subtract the exponents:** $\frac{x^a}{x^b} = x^{a-b}$.

3. **Raising a Power to a Power.** When simplifying an expression such as $(x^2)^3$, the quantity being raised to the third power is $x^2$, not just $x$. Think about this expression in expanded form. $(x^2)^3 = (x^2)(x^2)(x^2)$ which is equal to $x^6$. Using exponents of 2 and 3 resulted in an exponent of 6. In this type of situation, multiply the exponents.

 RULE BOOK

**When raising a power to another power, multiply the exponents:** $(x^a)^b = x^{a \cdot b}$.

4. **Raising a Product to a Power.** When there are multiple bases inside parentheses that are being raised to a power, each base in the parentheses must be evaluated with that power. For example, when simplifying the expression $(2x^2y^3)^3$, it becomes $2^3 x^{2 \cdot 3} y^{3 \cdot 3}$ which simplifies to $8x^6y^9$. The exponent outside the parentheses was evaluated on each base: 2, $x$, and $y$.

 RULE BOOK

**When raising a product to a power, raise each base to that power:** $(x^a y^b)^c = x^{a \cdot c} y^{b \cdot c}$

5. **Raising a Quotient to a Power.** If a quotient is raised to an exponent, evaluate both the numerator and denominator by the exponent. For example, $\left(\frac{x^3}{3}\right)^2 = \frac{x^{3 \cdot 2}}{3^2} = \frac{x^6}{9}$. Notice that the exponent of 2 belongs to both the numerator and the denominator of the fraction.

RULE BOOK

When raising a quotient to a power, raise both the numerator and denominator to that exponent; $\left(\frac{x}{y}\right)^a = \frac{x^a}{y^a}$.

## SOLVING EQUATIONS WITH EXPONENTS

An example of a question involving exponents in an equation may be as follows:

Solve for $x$: $2^{x+2} = 8^3$.
First, get the base numbers equal. Since 8 can be expressed as $2^3$, then $8^3 = (2^3)^3 = 2^9$. Both sides of the equation have a common base of 2, $2^{x+2} = 2^9$, so set the exponents equal to each other to solve for $x$; $x + 2 = 9$. So, $x = 7$.

### ▶ Negative Exponents

Any base number raised to a negative exponent is the *reciprocal* of the base raised to a positive exponent. For instance, $3^{-2} = \left(\frac{1}{3}\right)^2 = \frac{1}{9}$. Many questions will offer answer choices with only positive exponents, so remember to take the reciprocal when you need to make exponents positive. For example, the expression $x^{-3}$ can be rewritten as $\frac{1}{x^3}$ and the expression $\frac{1}{y^{-2}}$ can be rewritten as $y^2$.

 RULE BOOK

When simplifying with negative exponents, remember that $x^{-a} = \frac{1}{x^a}$.

### ▶ Fractional Exponents

When using fractional exponents, the numerator (top number) refers to the power of the base and the denominator (bottom number) refers to the root of the base. Any number to a fractional exponent with a numerator of 1 is the root of the number. Here are a few examples:

$25^{\frac{1}{2}} = \sqrt{25} = 5$ because $25 = 5^2 = 5 \cdot 5$.
$27^{\frac{1}{3}} = \sqrt[3]{27} = 3$ because $27 = 3^3 = 3 \cdot 3 \cdot 3$.
$64^{\frac{1}{4}} = \sqrt[4]{64} = 4$ because $64 = 4^4 = 4 \cdot 4 \cdot 4 \cdot 4$.

 RULE BOOK

A fractional exponent with a numerator of 1 denotes the root of the number: $x^{\frac{1}{a}} = \sqrt[a]{x}$.

When a number is raised to a fractional exponent where the numerator is not 1, take the root from the denominator and evaluate. Then raise the resulting base to the numerator. This is shown in the following example of 32 raised to the $\frac{2}{5}$ power. The value of $32^{\frac{2}{5}} = (\sqrt[5]{32})^2 = (2)^2 = 4$.

### EXTRA HELP

For further practice and explanation on the topic of exponents see Lesson 13: Working with Exponents from LearningExpress's *Algebra Success in 20 Minutes a Day.*

## ▶ *Radicals*

Taking the root of a number is the inverse, or opposite, of raising it to a power. A square root symbol is also known as a radical sign. The number inside the radical is the radicand. Taking the square root of a number is the same as raising that number to the $\frac{1}{2}$ power.

### EXTRA HELP

For some on-line practice with square roots, try the website www.aplusmath.com. Click on **Flashcards** and select **Square Roots**.

Many types of questions on certain tests will ask you to identify and perform operations with irrational numbers. Use the following definition to help relate this to your study of radicals.

### ●●●● GLOSSARY

**IRRATIONAL NUMBER** a number that cannot be written in fraction form. Irrational numbers are nonrepeating, nonterminating decimals. Any square root of a non-perfect square, such as $\sqrt{2}$ and $\sqrt{5}$, is an irrational number.

Recall that a perfect square is a number that is the result of multiplying two of the same integers. For instance, 36 is a perfect square because $6 \bullet 6 = 36$.

When taking the square root of a number, there are two answers: the positive root and the negative root. Note that $3 \bullet 3 = 9$, but also $-3 \bullet -3 = 9$. Therefore, when taking the square root of 9 the answer can be 3 or –3. So, how do you know which one to use in a particular question? If the radical reads $\sqrt{9}$ assume the positive root, and if the radical reads $-\sqrt{9}$ assume the negative root. If both roots are needed the radical will read $\pm\sqrt{9}$ which means "plus or minus the square root of 9" and would simplify to ±3. The solutions would be both 3 and –3 in this case.

Be careful when dealing with negative numbers and the radical sign.

Since two identical factors are needed to create a square, the square root of a negative number is not a real number. The only way to get a negative number as a product is to have one negative factor and one positive factor. When you evaluate the square root of a negative number on a calculator you will get an error message because it is only computing with real numbers.

### RULE BOOK

The square root of a negative number cannot be written as a real number; $\sqrt{-25}$ is an *imaginary number*.

When taking the cube root of a certain quantity you are looking for the number that when raised to the third power is equal to that quantity. Cube roots are written as $x^{\frac{1}{3}} = \sqrt[3]{x}$. Unlike square roots, the cube of a number can be negative because there are an odd number of factors. The cube root of 27 is 3. The cube root of –64 is –4.

## REDUCING RADICALS TO LOWEST RADICAL FORM

When reducing radicals, the objective is to get the number under the radical, or the *radicand*, as small as possible.

### GLOSSARY

**RADICAND** the value under the radical sign

To do this, find the largest perfect square factor of the radicand. The following are examples of perfect squares: 1, 4, 9, 16, 25, 36, 49, 64, 81, 100.

$$\sqrt{32} = \sqrt{16} \cdot \sqrt{2}$$

Take the square root of the perfect square and leave any remaining numbers under the radical; $\sqrt{32} = 4\sqrt{2}$.

Many times radicals will also involve variables. To reduce a radical with variables, remember that each pair of factors forms a perfect square. So, every 2 of the same variable equals one factor outside of the radical. For example, $\sqrt{x^2} = x$.

In a radical such as $\sqrt{x^3 y^5} = \sqrt{x \cdot x \cdot x \cdot y \cdot y \cdot y \cdot y \cdot y} = x \cdot y \cdot y \sqrt{x \cdot y} = xy^2\sqrt{xy}$.

## SHORTCUT

Any variable with an even exponent is a perfect square. To find the square root, divide the exponent by 2 and keep the base. For example, $\sqrt{x^4} = x^2$ and $\sqrt{y^6} = y^3$.

## COMBINING RADICALS

To add or subtract radicals, you must have *like terms*. In other words the radicand must be the same. If you have like terms, simply add or subtract the coefficients and keep the radicand the same.

*Examples:*
1. $3\sqrt{2} + 2\sqrt{2} = 5\sqrt{2}$
2. $4\sqrt{5} - \sqrt{5} = 4\sqrt{5} - 1\sqrt{5} = 3\sqrt{5}$
3. $6\sqrt{2} + 3\sqrt{5}$
   These cannot be combined because they are not *like terms* since the radicands, 2 and 5, are not the same.
4. $\sqrt{24} + 3\sqrt{6} = \sqrt{4} \cdot \sqrt{6} + 3\sqrt{6} = 2\sqrt{6} + 3\sqrt{6} = 5\sqrt{6}.$
   Notice in this example how $\sqrt{24}$ was reduced to $2\sqrt{6}$ before being added to $3\sqrt{6}$. This was to get them in the form of *like terms*.

## RULE BOOK

When adding and subtracting radicals, remember that $\sqrt{(x + y)} \neq \sqrt{x} + \sqrt{y}$ and $\sqrt{(x - y)} \neq \sqrt{x} - \sqrt{y}$. You may only combine *like terms* that have the same radicand.

## MULTIPLYING AND DIVIDING RADICALS

When multiplying and dividing radicals, you do *not* need like terms. When multiplying, simply multiply the numbers in front of the radicals and then the values inside the radicals, and keep the product of the radicals under the radical sign unless you can reduce.

RULE BOOK

**When multiplying radicals, simply multiply:**

$$\sqrt{x} \cdot \sqrt{y} = \sqrt{xy}$$

$$\sqrt{2} \cdot \sqrt{3} = \sqrt{6}$$

$$2\sqrt{3} \cdot 3\sqrt{5} = 6\sqrt{15}$$

When dividing, or simplifying a quotient, split up the numerator and denominator and reduce each separately from the other. Then divide the numbers in front of the radicals if possible and then the values inside the radicals. Note that a radical is not in simplest form if there is a fraction under the radical sign.

 RULE BOOK

**When dividing radicals, just divide:**

$$\sqrt{\frac{x}{y}} = \frac{\sqrt{x}}{\sqrt{y}}$$

$$\sqrt{\frac{25}{16}} = \frac{\sqrt{25}}{\sqrt{16}} = \frac{5}{4}$$

$$\frac{14\sqrt{10}}{7\sqrt{5}} = 2\sqrt{2}$$

Keep in mind that an expression is also not in simplest form if you leave a radical in the denominator. In order to remedy this situation, make the denominator into a perfect square. This step is called *rationalizing the denominator*. Here's how to do it:

Take the expression $\frac{6}{\sqrt{5}}$. This is reduced except for the fact that there is a radical in the denominator. To take care of this, multiply both the denominator and numerator by $\sqrt{5}$;  $\frac{6 \cdot \sqrt{5}}{\sqrt{5} \cdot \sqrt{5}}$. This results in the expression $\frac{6\sqrt{5}}{\sqrt{25}}$, which is equivalent to $\frac{6\sqrt{5}}{5}$. This is now in simplest form. Note that you cannot simplify the $\sqrt{5}$ and the whole number 5 any further.

 RULE BOOK

**A radical is reduced if:**

1. There are no perfect square factors under the radical.
2. There are no fractions under the radical.
3. There are no radicals in the denominator of an expression.

## CALCULATOR TIP

Many times questions using radicals can be made much easier by changing square roots to their decimal form, especially if the question is in a multiple-choice format. Use the radical (square root) button to find square roots. This is often the inverse or second function on the $x^2$ key. If a problem contains a root other than a square root, use the

or the

For example, to find the cube root of 64, press

   .

The result is 4.

## TIPS AND STRATEGIES

There are many strategies that are helpful when simplifying and evaluating exponents and radicals. Here are some helpful hints to use as you continue exploring these topics.

- Any base number or variable written without an exponent has an exponent of 1.
- A number raised to the zero power is equal to one.
- When multiplying like bases, add the exponents.
- When dividing like bases, subtract the exponents.
- When raising a product or quotient to a power, be sure to raise *all* bases to that power.
- Remember when squaring a number, there are two possible choices that result in the same square (i.e., $2^2 = 4$ and $(-2)^2 = 4$).
- When reducing radicals, find the *largest* perfect square factor of the radicand.
- On complicated algebra questions, substitute numbers to try to find an answer choice that is reasonable.
- Use the decimal form of radicals if possible, especially on multiple-choice exams.
- Remember that math is learned by *doing* problems—practice makes perfect!

## EXTRA HELP

**For further practice and explanation on reducing and performing operations with radicals see Lesson 18: Simplifying Radicals in Learning-Express's *Algebra Success in 20 Minutes a Day*.**

## CHAPTER QUIZ

The following questions provide practice on the skills needed to evaluate and simplify exponents and radicals. Check your progress on these topics with the answer explanations at the conclusion of this section.

1.  What is the numerical value of $3^5$?
    a. 8
    b. 15
    c. 27
    d. 81
    e. 243

2.  Which of the following is equivalent to $7 \bullet 7^5$?
    a. $7^2 \bullet 7^3$
    b. $7(7)^6$
    c. $7^6$
    d. $(7^3)^3$
    e. none of these

3.  Simplify the expression $\frac{4^4}{4^9}$.
    a. $4^5$
    b. $\frac{1-5}{4}$
    c. $4^{13}$
    d. $4^{36}$
    e. $\frac{1}{4^5}$

4.  Evaluate $x^2 + 3y - z$ when $x = 5$, $y = -3$, and $z = 4$.
    a. $-3$
    b. 30
    c. 15
    d. 12
    e. $-15$

5.  Evaluate $x(y^2 + 4) - 3$ when $x = 3$ and $y = -2$.
    a. $-3$
    b. 3
    c. 21
    d. $-24$
    e. $-23$

**6.** Simplify: $x^5 \bullet x^4$.
   **a.** $2x^9$
   **b.** $x^9$
   **c.** $x^{20}$
   **d.** $2x^{20}$
   **e.** $2x$

**7.** Simplify: $a^2b^3 \bullet ab^6$.
   **a.** $2a^3b^9$
   **b.** $a^2b^9$
   **c.** $a^3b^9$
   **d.** $a^2b^{18}$
   **e.** $2a^2b^3$

**8.** Simplify: $7y^2 \bullet 2xy$.
   **a.** $9xy^2$
   **b.** $9xy^3$
   **c.** $14xy^3$
   **d.** $14xy^2$
   **e.** $28xy^2$

**9.** Simplify: $\frac{x^{10}}{x^2}$.
   **a.** $x^8$
   **b.** $x^5$
   **c.** $x^{12}$
   **d.** $x^{20}$
   **e.** $2x^{20}$

**10.** Simplify: $\frac{3d^3}{9d^4}$.
   **a.** $3d$
   **b.** $3d^7$
   **c.** $\frac{1}{3d^7}$
   **d.** $\frac{1}{3d}$
   **e.** $27d^{12}$

**11.** Simplify: $\frac{2x^2 \bullet 4xy^2}{2x^2y}$.
   **a.** $4xy$
   **b.** $\frac{4y^2}{x}$
   **c.** $4y$
   **d.** $\frac{8y}{x}$
   **e.** $16xy$

**12.** Simplify: $(a^2b^3)^4$.
   **a.** $a^6b^7$
   **b.** $a^8b^{12}$
   **c.** $a^2b^{12}$
   **d.** $a^2b^7$
   **e.** $4a^2b^3$

**13.** Simplify: $6x^2(2x^3y)^4$.
   **a.** $8x^{14}y^4$
   **b.** $12x^{16}y^4$
   **c.** $192x^9y^4$
   **d.** $96x^{14}y^4$
   **e.** $24x^5y^4$

**14.** What is the value of $\left(\frac{x^6}{2}\right)^2$ in simplest form?
   **a.** $\frac{x^8}{2}$
   **b.** $\frac{x^{12}}{2}$
   **c.** $\frac{x^8}{4}$
   **d.** $\frac{x^{12}}{4}$
   **e.** $2x^{12}$

**15.** Which value of $x$ satisfies the equation $2^x = 32$?
   **a.** 2
   **b.** 3
   **c.** 4
   **d.** 5
   **e.** 16

**16.** In the equation $3^{2x+2} = 81$, what is the value of $x$?
   **a.** 0
   **b.** 1
   **c.** 2
   **d.** 9
   **e.** 81

**17.** Which of the following is equivalent to $2^3a^{-3}$?
   **a.** $6a^3$
   **b.** $8a^3$
   **c.** $\frac{6}{a^3}$
   **d.** $\frac{8}{a^3}$
   **e.** $\frac{8}{a^3}$

EXPONENTS

**18.** Which of the following is equivalent to $ab^{-3}c^2$?

   **a.** $ab^3c^2$

   **b.** $ab^3c^{-2}$

   **c.** $\frac{ab^3}{c^2}$

   **d.** $\frac{ac^2}{b^3}$

   **e.** $\frac{b^3}{ac^2}$

**19.** What is the value of $\sqrt{96}$ in simplest radical form?

   **a.** $9\sqrt{6}$

   **b.** $2\sqrt{24}$

   **c.** $4\sqrt{6}$

   **d.** $16\sqrt{6}$

   **e.** $48\sqrt{2}$

**20.** What is the value of $(\sqrt[3]{216})^2$?

   **a.** 6

   **b.** 12

   **c.** 36

   **d.** 144

   **e.** 324

**21.** What is the value of $\sqrt{24x^3}$ in simplest radical form?

   **a.** $2x\sqrt{6}$

   **b.** $2x\sqrt{6x}$

   **c.** $6x\sqrt{2}$

   **d.** $6x\sqrt{2x}$

   **e.** $24x\sqrt{x}$

**22.** Which of the following is equal to $64^{\frac{2}{3}}$?

   **a.** 4

   **b.** 8

   **c.** 16

   **d.** 32

   **e.** 512

**23.** Which of the following statements is NOT true?

   **a.** $\sqrt{2} + 2\sqrt{2} = 3\sqrt{2}$

   **b.** $3\sqrt{2} + 2\sqrt{3} = 5\sqrt{5}$

   **c.** $7\sqrt{2} - \sqrt{8} = 5\sqrt{2}$

   **d.** $\sqrt{36} - \sqrt{25} = 1$

   **e.** $2\sqrt{3} + 4\sqrt{3} = 6\sqrt{3}$

**24.** Which of the following represents the product of $8\sqrt{3}$ and $\sqrt{20}$ in simplest radical form?
   **a.** $8\sqrt{60}$
   **b.** $16\sqrt{8}$
   **c.** $16\sqrt{15}$
   **d.** $32\sqrt{15}$
   **e.** $\sqrt{480}$

**25.** Which of the following is equivalent to $\frac{4\sqrt{6}}{2\sqrt{2}}$?
   **a.** $2\sqrt{3}$
   **b.** $2\sqrt{12}$
   **c.** $8\sqrt{3}$
   **d.** $8$
   **e.** $8\sqrt{12}$

## ANSWERS

Using the answer explanations to the quiz can be an important tool in finding errors and misconceptions in the material. Carefully check your solutions with the corresponding answers to see how you are doing with exponents and radicals.

**1. e.** The expression $3^5$ is equivalent to $3 \bullet 3 \bullet 3 \bullet 3 \bullet 3$. When multiplied, this is equal to 243.

**2. c.** When multiplying like bases, add the exponents; $7 \bullet 7^5$ can also be written as $7^1 \bullet 7^5$. Add the exponents to get $7^{1+5}$ which equals $7^6$.

**3. e.** When dividing like bases, subtract the exponents. The expression $\frac{4^4}{4^9}$ becomes $4^{4-9}$ which is equal to $4^{-5}$. To write this with a positive exponent, take the reciprocal of the base. The final answer is $\frac{1}{4^5}$.

**4. d.** The first step is to substitute the values of each variable into the expression; $(5)^2 + 3(-3) - 4$. Order of operations says do exponents next; $25 + 3(-3) - 4$. Multiply $3 \bullet -3$; $25 + -9 - 4$. Do addition and subtraction in order from left to right; $25 + -9 - 4$; $16 - 4 = 12$.

**5. c.** Substitute the values for the variables in the expression; $(3)((-2)^2 + 4) - 3$. Evaluate the exponent inside the parentheses; $(3)((-2)^2 + 4) - 3$. Remember that $(-2)^2 = (-2)(-2) = 4$. Add in-side the parentheses; $(3)(4 + 4) - 3$. Multiply the first term; $(3)(8) - 3$. Subtract; $24 - 3 = 21$.

**6. b.** When multiplying like bases, add the exponents; $x^{5+4} = x^9$.

**7. c.** Remember, anytime a variable does not have an exponent, assume the exponent is 1. In this case, $a = a^1$. Use commutative property to arrange same letters next to each other. $a^2 a \cdot b^3 b^6$. When multiplying like bases, add the exponents; $a^{2+1} \cdot b^{3+6} = a^3 b^9$.

**8. c.** Use commutative property to arrange like variables and the coefficients next to each other; $7 \cdot 2 \cdot x \cdot y^2 \cdot y$. Multiply. Don't forget to add the exponents of the like bases. The result is $14xy^3$.

**9. a.** When dividing like bases, subtract the exponents; $x^{10-2} = x^8$.

**10. d.** Reduce the coefficients; $\frac{1d^3}{3d^4}$. When dividing like bases, subtract the exponents; $\frac{1d^{3-4}}{3}$. Simplify; $\frac{1d^{-1}}{3}$. A variable in the numerator with a negative exponent is equal to the same variable in the denominator with the opposite (positive) sign on the exponent. The answer is $\frac{1}{3d^1}$ which is equal to $\frac{1}{3d}$.

**11. a.** Use commutative property in the numerator to arrange like variables and the coefficients together; $\frac{2 \cdot 4x^2 \cdot xy^2}{2x^2 y}$. Multiply in the numerator. Remember that $x^2 \cdot x = x^2 \cdot x^1 = x^{2+1} = x^3$. The result is $\frac{8x^3 y^2}{2x^2 y}$. Divide the coefficients; $8 \div 2 = 4$ so the expression becomes $\frac{4x^3 y^2}{x^2 y}$. Divide the variables by subtracting the exponents; $4x^{3-2}y^{2-1}$. Simplify; $4x^1 y^1 = 4xy$.

**12. b.** Multiply the outer exponent by each exponent inside the parentheses; $a^{2 \cdot 4} b^{3 \cdot 4}$. Simplify to get a result of $a^8 b^{12}$.

**13. d.** Evaluate the exponent by multiplying each number or variable inside the parentheses by the exponent outside the parentheses; $6x^2(2^4 x^{3 \cdot 4} y^4)$. Simplify; $6x^2(16x^{12}y^4)$. Multiply the coefficients and add the exponents of like variables; $6 \cdot 16x^{2+12}y^4$. Simplify to get a result of $96x^{14}y^4$.

**14. d.** Raise both the numerator and denominator to the power of 2. The expression $\left(\frac{x^6}{2}\right)^2$ then becomes $\frac{x^{6 \cdot 2}}{2^2}$ which simplifies to $\frac{x^{12}}{4}$.

**15. d.** Since 32 can be written as $2^5$, the equation becomes $2^x = 2^5$. Therefore, $x$ must equal 5.

**16. b.** Since 81 can be written as $3^4$, the equation becomes $3^{2x+2} = 3^4$. Since the bases are equal, set the exponents equal and solve for $x$;

$2x + 2 = 4$. Subtract 2 from both sides of the equal sign; $2x + 2 - 2 = 4 - 2$. This simplifies to $2x = 2$. Therefore, $x = 1$.

**17. e.** Evaluating $2^3$ results in 8 and the expression $a^{-3}$ written with a positive exponent becomes $\frac{1}{a^3}$. Therefore, the expression simplifies to $\frac{8}{a^3}$.

**18. d.** Write the expression with positive exponents by taking the reciprocal of any base with a negative exponent. The expression $ab^{-3}c^2$ then becomes $\frac{ac^2}{b^3}$. Bases that already had a positive exponent remain in the numerator.

**19. c.** To reduce radicals, find the largest perfect square factor of the radicand. Since 96 is equal to $16 \bullet 6$, then $\sqrt{96}$ can be expressed as $\sqrt{16} \bullet \sqrt{6}$. The square root of 16 is equal to 4, so the reduced radical becomes $4\sqrt{6}$.

**20. c.** To find the value of $(\sqrt[3]{216})^2$, start inside the parentheses and take the cube root of 216. This is equal to 6. Now raise that value of 6 to the second power; $6^2 = 36$.

**21. b.** First reduce the numerical part of the radical. Since 24 can be written as $4 \bullet 6$ then $\sqrt{24}$ can be written as $\sqrt{4} \bullet \sqrt{6}$ which is equal to $2\sqrt{6}$. The variable portion of the radical can be written as $\sqrt{x \bullet x \bullet x}$. Each pair of $x$'s inside the radical represents a perfect square so this can be reduced to $x\sqrt{x}$. Putting both parts together results in the reduced radical $2x\sqrt{6x}$.

**22. c.** An exponent of $\frac{2}{3}$ tells you to take the cube root of the base and then square the result. In a base of 64, the cube root is 4 because $4 \bullet 4 \bullet 4 = 64$. Following this, $4^2$ is 16 which is the final answer.

**23. b.** When combining like terms using radicals, the radicand (number under the radical) must be the same. Answer choice **b** is adding two numbers that do not have the same radicand, so choice **b** is not true. Each of the other choices is true.

**24. c.** *Product* is a key word for multiplication, so multiply the numbers in front of the radicals together and then the numbers inside the radicals; $8\sqrt{3} \bullet \sqrt{20}$ becomes $8 \bullet 1\sqrt{3} \bullet \sqrt{20}$. Simplify to get $8\sqrt{60}$. Since this is not an answer choice, reduce $\sqrt{60}$ to $\sqrt{4} \bullet \sqrt{15}$ which becomes $2\sqrt{15}$. Therefore, $8\sqrt{60}$ is equal to $8 \bullet 2\sqrt{15}$ which simplifies to $16\sqrt{15}$.

**25. a.** A radical is not in simplest form if there is a radical in the denominator. To eliminate this radical, multiply both numerator and denominator by $\sqrt{2}$; $\frac{4\sqrt{6} \cdot \sqrt{2}}{2\sqrt{2} \cdot \sqrt{2}}$ which becomes $\frac{4\sqrt{12}}{2\sqrt{4}}$. Since $\sqrt{12}$ reduces to $2\sqrt{3}$, the numerator becomes $4 \cdot 2\sqrt{3}$ which is $8\sqrt{3}$. Since $\sqrt{4}$ reduces to 2, the denominator becomes $2 \cdot 2$ which is 4. The fraction is $\frac{8\sqrt{3}}{4}$ which simplifies to $2\sqrt{3}$. Another way to solve this question is to simply divide the coefficients and then divide the radicals; $4 \div 2 = 2$ and $\sqrt{6} \div \sqrt{2} = \sqrt{3}$. This also results in $2\sqrt{3}$.

# Polynomials

Knowing how to work with polynomials is one of the basic foundations for learning and understanding algebra. As you start your review of this crucial topic, take a few minutes to take this ten-question *Benchmark Quiz*. These questions will cover many of the topics you are expected to know and are similar to what may appear on exams. Check the answer key carefully to monitor your progress and understanding. Your Benchmark Quiz analysis will help you determine if you need more practice with polynomials and what specific skills to spend that precious time on.

## BENCHMARK QUIZ

1. What is the degree of the monomial $6x^2y$?
   a. 0
   b. 1
   c. 2
   d. 3
   e. 6

**2.** Which of the following polynomials has degree 5?
   **a.** $5x - 5$
   **b.** $5x^2y + 4x - 5$
   **c.** $x^5y + xy^5$
   **d.** $x^4y + 3$
   **e.** none of these

**3.** Simplify: $2c + 4 + 5c - 1$.
   **a.** $7c + 3$
   **b.** $7c + 5$
   **c.** $3c + 3$
   **d.** $10c$
   **e.** $10c^2 + 3$

**4.** Simplify: $(x - 1) - (2x - 4)$.
   **a.** $-x - 5$
   **b.** $-x + 5$
   **c.** $x - 3$
   **d.** $-3x - 5$
   **e.** $-x + 3$

**5.** Simplify: $(3x^2)(-4x^6)$.
   **a.** $-7x^8$
   **b.** $-7x^{12}$
   **c.** $-12x^{12}$
   **d.** $-12x^8$
   **e.** $-x^8$

**6.** Simplify: $6(2x - y)$.
   **a.** $8xy$
   **b.** $8x - y$
   **c.** $8x - 6y$
   **d.** $12x - y$
   **e.** $12x - 6y$

**7.** Multiply: $cd(c^2 + 2cd - 1)$.
   **a.** $c^2d + 2cd - 1$
   **b.** $c^2d + 3cd - 1$
   **c.** $c^2d + 2c^2d^2 - cd$
   **d.** $c^3d + 2c^2d^2 - 1$
   **e.** $c^3d + 2c^2d^2 - cd$

**8.** Multiply the binomials: $(x + 2)(x + 3)$.
  **a.** $x^2 + 5x + 6$
  **b.** $x^2 + 5x + 5$
  **c.** $x^2 + 6x + 5$
  **d.** $x^2 + 2x + 6$
  **e.** $x^2 + 6$

**9.** Simplify: $(3x - 2)^2$.
  **a.** $9x^2 - 4$
  **b.** $9x^2 - 12x + 4$
  **c.** $9x^2 - 4x + 4$
  **d.** $3x^2 - 12x - 4$
  **e.** $3x^2 + 5x + 5$

**10.** Multiply the polynomials: $(p + 2)(p^2 + 2p + 1)$.
  **a.** $p^2 + 4p + 2$
  **b.** $p^3 + 2p^2 + 4p + 2$
  **c.** $p^3 + 2p^2 + 5p + 2$
  **d.** $p^3 + 4p^2 + 5p + 1$
  **e.** $p^3 + 4p^2 + 5p + 2$

## BENCHMARK QUIZ SOLUTIONS

How did you do on simplifying polynomials? Check your answers here, and then analyze the results to figure out your plan of attack to master these topics.

### ▶ Answers

**1. d.** The degree of a monomial is the sum of the exponents of the variables. The exponent on $x$ is 2 and the exponent on $y$ is 1; $2 + 1 = 3$. The degree is 3.

**2. d.** The degree of a polynomial is the degree of the term with the greatest degree. Add the exponents of the variables in each term and find the largest degree. Choice **a** has degree 1. Choice **b** has degree 3 because the first term has degree 3. Choice **c** has degree 6; each of the terms has degree 6. Choice **d** has degree 5. Add the exponents of $x$ and $y$ in the first term; $4 + 1$ equals 5.

**3. a.** Use the commutative property to place *like terms* next to each other; $2c + 5c + 4 - 1$. Combine like terms; $7c + 3$.

**4. e.** Eliminate the parentheses by distributing the subtraction to both terms in the second set of parentheses. The expression becomes $x - 1 - 2x + 4$. Use commutative property to place *like terms* next to each other; $x - 2x - 1 + 4$. Combine like terms; $-x + 3$.

**5. d.** Multiply coefficients and then add the exponents of any like bases; $(3x^2)(-4x^6)$ equals $3 \bullet -4x^{2+6}$. Simplify to get $-12x^8$.

**6. e.** Use the distributive property to multiply 6 by both terms inside the parentheses; $6(2x - y)$ becomes $6 \bullet 2x - 6 \bullet y$. This simplifies to $12x - 6y$.

**7. e.** Use the distributive property to multiply $cd$ by each of the terms in the parentheses; $cd(c^2 + 2cd - 1)$ becomes $(cd \bullet c^2) + (cd \bullet 2cd) - (cd \bullet 1)$. Multiply each term to simplify. Add the exponents of any like bases; $c^3d + 2c^2d^2 - cd$.

**8. a.** Use the distributive property (FOIL) to multiply each term in the first binomial by each term of the second; $(x \bullet x) + (3 \bullet x) + (2 \bullet x) + (2 \bullet 3)$. Simplify each term; $x^2 + 3x + 2x + 6$. Combine like terms; $x^2 + 5x + 6$.

**9. b.** Since a binomial is being raised to the second power, write the binomial as two factors. Then multiply using distributive property (FOIL); $(3x - 2)^2 = (3x - 2)(3x - 2)$; $(3x \bullet 3x) - (3x \bullet 2) - (2 \bullet 3x) + (-2 \bullet -2)$. Simplify each term; $9x^2 - 6x - 6x + 4$. Combine like terms; $9x^2 - 12x + 4$.

**10. e.** Use the distributive property to multiply the first term of the binomial, $p$, by each term of the trinomial, and then the second term of the binomial, 2, by each term of the trinomial; $(p + 2)(p^2 + 2p + 1) = (p \bullet p^2) + (p \bullet 2p) + (p \bullet 1) + (2 \bullet p^2) + (2 \bullet 2p) + (2 \bullet 1)$. Simplify by multiplying in each term; $p^3 + 2p^2 + 1p + 2p^2 + 4p + 2$. Use commutative property to arrange like terms next to each other; $p^3 + 2p^2 + 2p^2 + 1p + 4p + 2$. Combine like terms; $p^3 + 4p^2 + 5p + 2$.

## BENCHMARK RESULTS

If you answered 8–10 questions correctly, you have a solid foundation of operations with polynomials. Examine the lesson and focus on ideas that you need to review. Take the quiz at the end of the chapter to be sure that you have a good understanding of each of the concepts.

If you answered 4–7 questions correctly, you need to review a number of concepts that pertain to the material in this section. Read through the chapter carefully for a refresher and pay careful attention to the sidebars that refer you to more in-depth practice, hints, and shortcuts. Work through the quiz at the end of the chapter to check your progress.

If you answered 1–3 questions correctly, there are many topics that would be beneficial to review. Read the section, with the examples and sidebars, and concentrate on any skills that appear unfamiliar. Many of the skills in the chapter are those that appear in most high school algebra courses but they are forgotten if not used on a regular basis. The quiz at the end of the chapter will provide even more review and practice, so be sure to use the answer explanations for help. You may also want to reference a more in-depth and comprehensive book, such as LearningExpress's *Algebra Success in 20 Minutes a Day*.

## JUST IN TIME LESSON—POLYNOMIALS

In this lesson, the skills needed to perform operations on polynomials will be explained and demonstrated. Specifically, this section will cover:

- degree of polynomial terms
- combining like terms of polynomials
- multiplying various types of polynomials

### ▶ Degree of Polynomials

In algebra, you are using a letter to represent an unknown quantity. This letter is called the *variable*. The number preceding the variable is called the *coefficient*. If there is no number written in front of the variable, the coefficient is understood to be 1. When a coefficient or variable is raised to a power, this number is the *exponent*. Take a look at the following examples:

$5x$     Five is the coefficient, $x$ is the variable, 1 is the exponent.

$ab$     1 is the coefficient and both $a$ and $b$ are the variables and both have an exponent of 1.

$-4x^3y$     Negative four is the coefficient, $x$ and $y$ are the variables. 3 is the exponent on $x$ and 1 is the exponent on $y$.

Since different terms of polynomials are separated by addition and subtraction, each of the above examples represents a monomial. Even though there may be more than one number and/or variable involved, they are still monomials. Another important aspect of monomials is the degree of the monomial.

## GLOSSARY

**DEGREE OF A MONOMIAL** the sum of the exponents of the variables. The degree of the term $3x$ is 1 because $3x = 3x^1$. The degree of the term $x^2y^3$ is 5 because $2 + 3 = 5$.

## ▶ Combining Like Terms

A crucial algebraic concept to recognize is *like terms*. In algebra, *like terms* are expressions that have exactly the same variable(s) and exponents and can be combined easily by adding or subtracting the coefficients.

*Examples:*

| | |
|---|---|
| $4x + 5x$ | These are *like terms*, and the sum is $9x$. |
| $6x^2y + -11x^2y$ | These are also *like terms*, and the sum is $-5x^2y$. |
| $3a - (5a)$ | These are like terms, and the difference is $-2a$. |
| $9xy^2 + 9x^2y$ | These are NOT *like terms* because the exponents of the variables are not exactly the same. They cannot be combined. |

When subtracting polynomials in particular, it is imperative that the distributive property is used on each term in the polynomial being subtracted. Take the situation $(3x^2 - 4x - 5) - (2x^2 - 7x + 9)$. The subtraction sign in front of the polynomial $(2x^2 - 7x + 9)$ is treated as a $-1$. To eliminate the parentheses, each term individually will be changed to its opposite as the subtraction is distributed to the $2x^2$, $-7x$, and 9. The entire expression then becomes $3x^2 - 4x - 5 - 2x^2 + 7x - 9$. Complete the problem by combining *like terms*. Therefore the simplified expression is $x^2 + 3x - 14$.

When like terms cannot be combined they create an expression called a polynomial.

## GLOSSARY

**POLYNOMIAL** the sum or difference of two or more monomials. Terms of polynomials are separated by addition and subtraction.

Some polynomials have specific names:

| | |
|---|---|
| $3x^2$ | is a *monomial* because there is one term. |
| $7x + 8y$ | is a *binomial* because there are two terms. |
| $2x^2 + 6x - 7$ | is a *trinomial* because there are three terms. |

We have already learned how to find the degree of a monomial. We can also find the degree of different types of polynomials.

GLOSSARY

**DEGREE** of a polynomial is equal to the degree of the term of the polynomial with the greatest degree.

The degree of the polynomial $3x^2 + 5x - 9$ is 2. The first term, $3x^2$, has degree 2 because the exponent on the variable is 2. This is the term with greatest degree of this polynomial. The degree of the polynomial $5x^2y + 3xy - 2y$ is 3. The leading term of $5x^2y$ has an exponent of 2 on $x$ and 1 on $y$; 2 + 1 = 3. The second term only has degree 2 and the third term only has degree 1.

 EXTRA HELP

For additional examples and explanations, look to Learning-Express's *Algebra Success in 20 Minutes a Day* Lesson 3: Combining Like Terms.

## ▶ *Multiplying Polynomials*

When multiplying by a monomial, use the distributive property to simplify if there is more than one term to be multiplied. Multiply coefficients by coefficients, and add the exponents of any like bases.

*Examples:*
Multiply each of the following:
1. $(6x^3)(5xy^2) = 30x^4y^2$      (Remember that $x = x^1$.)
2. $2x(x^2 - 3) = 2x^3 - 6x$      Use distributive property.
3. $x^3(3x^2 + 4x - 2) = 3x^5 + 4x^4 - 2x^3$   Use distributive property.

When multiplying two binomials together, use an acronym called FOIL.

 SHORTCUT

**FOIL** is a way to remember how to multiply two binomials using the distributive property.

    **F** Multiply the first terms in each set of parentheses.

    **O** Multiply the outer terms in the parentheses.

    **I** Multiply the inner terms in the parentheses.

    **L** Multiply the last terms in the parentheses.

The following examples show different cases of using FOIL to multiply two binomials.

*Examples:*

1.

   $= x^2 + 2x - 1x - 2 = x^2 + x - 2$
   F O I L

2. $(y - 3)^2 = (y - 3)(y - 3) = y^2 - 3y - 3y + 3^2 = y^2 - 2 \cdot 3y + 3^2 =$
   $y^2 - 6y + 9$.          F O I L

The previous model is an example of a common situation on many standardized tests. It is an example of the square of a binomial difference. Another common situation is the square of a binomial sum.

 SHORTCUT

**When finding the square of a binomial difference, use the formula**
$(x - y)^2 = x^2 - 2xy + y^2$.
**When finding the square of a binomial sum, use the formula** $(x + y)^2 = x^2$
$+ 2xy + y^2$.

In addition to these examples, another common type of binomial multiplication problem is finding the product of the sum and the difference of the same two values. The general form is $(x - y)(x + y)$ and will multiply to the binomial $x^2 - y^2$.

SHORTCUT

**When multiplying the sum and difference of the same two terms the result is** $x^2 - y^2$.

When multiplying any type of polynomial by another polynomial, the distributive property should always be used. Take, for example, the binomial $x + 2$ and the trinomial $2x^2 - 3x - 5$. In order to multiply these together, you must multiply the $x$ from the binomial by each term of the trinomial. Then repeat the process with the 2 from the binomial. This is what the process would look like. To multiply $(x + 2)(2x^2 - 3x - 5)$: Multiply each term of the trinomial first by $x$ and then by 2: $(2x^2 \cdot x) - (3x \cdot x) - (5 \cdot x) + (2x^2 \cdot 2) - (3x \cdot 2) - (5 \cdot 2)$. Now simplify each term; $2x^3 - 3x^2 - 5x + 4x^2 - 6x - 10$. Combine *like terms* to get the simplified expression; $2x^3 + x^2 - 11x - 10$.

## EXTRA HELP

For additional examples and explanations, look to the website www. mathforum.org. From the Home page under **Mathematics Topics,** click on **Algebra.** Choose **Basic Algebra** and then **Polynomials.** This page will offer links for algebra resources in this category and many others. You can find additional assistance in book form in Lesson 14: Polynomials in LearningExpress's *Algebra in 20 Minutes a Day.*

## TIPS AND STRATEGIES

Use the following summary to highlight the facts and skills of the chapter and make your plan to conquer polynomials.

- The degree of a monomial is the sum of the exponents of the variables.
- The degree of a polynomial is the same as the term with the greatest degree.
- Terms of polynomials are separated by addition and subtraction.
- Like terms are terms that have *exactly* the same variables and exponents.
- Be careful to distribute the negative sign when subtracting polynomials.
- Only concern yourself with *like terms* when adding or subtracting polynomials. Like terms are not necessary for multiplication and division.
- Use the acronym FOIL to assist in multiplying two binomials.
- Use the formula $(x - y)^2 = x^2 - 2xy + y^2$ when squaring a binomial difference and use $(x + y)^2 = x^2 + 2xy + y^2$ for a binomial sum.
- Remember that $(x - y)(x + y) = x^2 - y^2$.
- Use distributive property to multiply in all other situations with polynomials.
- On multiple-choice tests it is often best to work out the problem first before looking at the answer choices. Remember that the wrong answer choices are often the result of making common errors.
- Keep in mind that new mathematics topics build on previous knowledge. Master the basics and that will carry you through the more difficult concepts to come.

## CHAPTER QUIZ

Try these practice problems as you track your progress through simplifying polynomials.

1. What is the degree of the monomial $10x^3y^2$?
   a. 2
   b. 3
   c. 5
   d. 6
   e. 10

2. Which of the following monomials has degree 3?
   a. $3x$
   b. $3x^2$
   c. $3xy$
   d. $3x^2y$
   e. $3x^3y$

3. What is the degree of the polynomial $5xy^2 + 3x$?
   a. 2
   b. 3
   c. 5
   d. 8
   e. 12

4. Simplify the expression $-4y - 3y + 2y$.
   a. $-5y$
   b. $5y$
   c. $y$
   d. $-3y$
   e. $-9y$

5. Simplify the expression $6a + 2 - 2a$.
   a. $a$
   b. $a + 2$
   c. $4a + 2$
   d. $6a$
   e. $8a + 2$

6. Simplify the expression: $2a^2b + 4a^2b + a^2b$.
   a. $7a^2b$
   b. $7a^6b^3$
   c. $6a^2b$
   d. $6a^6b^3$
   e. $8a^6b^3$

**7.** Simplify the expression: $3xy^2 + 4x^2y - xy^2$.
   **a.** $7xy^2$
   **b.** $6xy^2$
   **c.** $7x^2y^2 - xy^2$
   **d.** $2xy^2 + 4x^2y$
   **e.** $6x^2y^2$

**8.** Which of the following choices shows the expression $7(x - 3) + 21$ in simplified form?
   **a.** $7x$
   **b.** $7x + 18$
   **c.** $10x + 21$
   **d.** $10x$
   **e.** $7x + 24$

**9.** Simplify: $(2a - b) + (7a + b)$.
   **a.** $9a$
   **b.** $9a + b$
   **c.** $9a - b$
   **d.** $14a$
   **e.** $14a - b$

**10.** Simplify the expression: $9(2x - 1) - 2(x - 5)$.
   **a.** $16x - 6$
   **b.** $16x + 1$
   **c.** $20x - 6$
   **d.** $16x - 1$
   **e.** $9x - 6$

**11.** Subtract $(3x^2 - 4x + 9)$ from $(9x^2 - 6x + 1)$.
   **a.** $-6x^2 - 10x + 8$
   **b.** $6x^2 - 10x - 8$
   **c.** $6x^2 - 2x - 8$
   **d.** $6x^2 - 2x + 8$
   **e.** $12x^2 - 10x - 8$

**12.** Simplify the expression: $\frac{3}{5}(15x + 10y)$.
   **a.** $9x + 10y$
   **b.** $9x + 6y$
   **c.** $6x + 5y$
   **d.** $6x + 6y$
   **e.** $45x + 2y$

**13.** Multiply: $2x(7x + 5)$.
   **a.** $19x$
   **b.** $19x^2$
   **c.** $14x^2 + 5x$
   **d.** $14x^2 + 10x$
   **e.** $14x + 10$

**14.** If the length of a rectangle is expressed as $4y$ and the width as $5y + 8$, what is the area of the rectangle in terms of $y$?
   **a.** $9y$
   **b.** $9y + 8$
   **c.** $18y + 16$
   **d.** $20y + 12$
   **e.** $20y^2 + 32y$

**15.** Simplify the expression: $(3x)(4y) - (6x)(3y)$.
   **a.** $-6xy$
   **b.** $6xy$
   **c.** $18xy$
   **d.** $2xy$
   **e.** $-2xy$

**16.** Multiply: $-8p^3r(2p - 3r)$.
   **a.** $-10p^4r + 11p^3r^2$
   **b.** $-16p^4r + 24p^3r^2$
   **c.** $-16p^4r - 5r$
   **d.** $-16p^3r + 24p^3r^2$
   **e.** $16p^3r - 24p^3r$

**17.** Multiply the binomials: $(x + 2)(x + 4)$.
   **a.** $x^2 + 6$
   **b.** $x^2 + 8$
   **c.** $x^2 + 6x + 8$
   **d.** $x^2 + x + 6$
   **e.** $2x + 8$

**18.** What is the product of the binomials $(2x - 1)$ and $(x + 8)$?
   **a.** $2x^2 + 15x - 8$
   **b.** $2x^2 - 8$
   **c.** $2x^2 - 15x - 8$
   **d.** $2x^2 + 17x - 8$
   **e.** $2x - 8$

**19.** What is the product of $(2x - 1)$ and $(2x + 1)$?
   **a.** $2x^2 - 1$
   **b.** $4x^2 + 1$
   **c.** $4x^2 - 1$
   **d.** $4x^2 + 4x + 1$
   **e.** $4x^2 + 4x - 1$

**20.** Simplify the expression $(x - 1)^2$.
   **a.** $x^2 - 1$
   **b.** $x^2 - 2$
   **c.** $x^2 - x + 1$
   **d.** $x^2 + x - 1$
   **e.** $x^2 - 2x + 1$

**21.** Multiply the binomials: $(4x^2 - 2)(3x - 8)$.
   **a.** $12x^2 - 6x + 16$
   **b.** $12x^3 - 32x^2 - 6x + 16$
   **c.** $12x^2 - 32x^2 - 6x + 16$
   **d.** $12x^3 + 32x^2 - 6x - 16$
   **e.** $12x^3 - 32x^2 - 5x + 16$

**22.** Multiply the binomial by the trinomial: $(x - 3)(x^2 + 4x + 2)$.
   **a.** $x^3 + x^2 - 12x - 6$
   **b.** $x^3 + 7x^2 - 12x - 6$
   **c.** $x^3 - x^2 - 10x - 6$
   **d.** $x^3 + x^2 - 10x - 6$
   **e.** $x^2 - 10x - 6$

**23.** Which of the following represents the product of $(2x - 1)$ and $(x^2 - 3x - 4)$?
   **a.** $x^2 + 5x - 1$
   **b.** $2x^2 - 3x + 4$
   **c.** $2x^3 - 7x^2 - 5x + 4$
   **d.** $2x^3 - 7x^2 - 5x - 5$
   **e.** $2x^3 - 5x - 4$

**24.** Multiply the binomials: $(x + 1)(x - 5)(x + 4)$.
   **a.** $x^3 - 20$
   **b.** $x^3 + 8x^2 + 21x - 20$
   **c.** $x^3 - 8x^2 - 21x - 20$
   **d.** $x^3 - 21x + 20$
   **e.** $x^3 - 21x - 20$

**25.** If the dimensions of a rectangular prism are expressed as $x - 3$, $x + 4$, and $2x + 1$ what is the volume of the prism in terms of $x$?

    **a.** $x^3 - 12$

    **b.** $2x^3 + 3x^2 - 25x - 12$

    **c.** $2x^3 + 3x^2 - 23x - 12$

    **d.** $2x^3 + 2x^2 - 25x - 12$

    **e.** $2x^3 - 23x - 12$

## ANSWERS

Use the explanations written here to help spell out misconceptions and errors. You can also reference the website mentioned in the text, or the other comprehensive texts cited for extra help on polynomials.

**1. c.** The degree of a monomial is the sum of the exponents on the variables. The exponent on $x$ is 3 and the exponent on $y$ is 2; $3 + 2 = 5$. The degree is 5.

**2. d.** The degree is the sum of the exponents on the variables. Choice **a** has degree 1. Choice **b** has degree 2. Choice **c** has degree 2. Choice **d** has degree 3. The exponents are 2 and 1; $2 + 1 = 3$. The degree of choice **e** is 4.

**3. b.** The degree of a polynomial is the degree of the term with the greatest degree. The first term in the polynomial has degree 3 because the exponents are 1 and 2. The second term has degree 1. The term with the greatest degree has degree of 3, so the polynomial has degree 3.

**4. a.** Add and subtract like terms in order from left to right; $-4y - 3y + 2y$. Change subtraction to addition and the sign of the following term to its opposite; $-4y + - 3y + 2y$. The signs are different so subtract the coefficients; $-7y + 2y = -5y$.

**5. c.** Use commutative property to arrange like terms together; $6a - 2a + 2$. Subtract like terms; $6a - 2a + 2 = 4a + 2$.

**6. a.** Each one of the terms are like terms because the variables in each are exactly the same, including the exponents. To simplify the expression, combine the coefficients; $2 + 4 + 1 = 7$. Keep the variables that were originally with each term to make $7a^2b$.

**7. d.** Because not every term has exactly the same variable and exponent configuration, you cannot just combine the coefficients here. You may only combine like terms, which are the first and third term; $3xy^2 + 4x^2y - xy^2$. Use commutative property to put together like terms; $3xy^2 - xy^2 + 4x^2y$. Subtract like terms; $2xy^2 + 4x^2y$.

**8. a.** Eliminate the parentheses first by using the distributive property; $7(x - 3) + 21$ becomes $7x - 21 + 21$. Combine like terms; $7x + 0 = 7x$.

**9. a.** Use commutative property of addition to group together *like terms*; $2a + 7a - b + b$. Combine like terms and the solution is $9a$. Notice that $-b + b = 0b = 0$.

**10. b.** Change subtraction signs to addition and change the sign of the following number to its opposite; $9(2x + -1) + -2(x + -5)$. Eliminate the both sets of parentheses first by using the distributive property; $18x + -9 + -2x + 10$. Use commutative property of addition to group together like terms; $(18x + -2x) + (-9 + 10)$. Combine like terms. The solution is $16x + 1$.

**11. c.** To subtract $(3x^2 - 4x + 9)$ from $(9x^2 - 6x + 1)$ the expression becomes $(9x^2 - 6x + 1) - (3x^2 - 4x + 9)$. Eliminate the parentheses by using distributive property; $9x^2 - 6x + 1 - 3x^2 + 4x - 9$. Use commutative property of addition to group together like terms; $(9x^2 - 3x^2) + (-6x + 4x) + (1 - 9)$. Combine like terms; $6x^2 - 2x - 8$.

**12. b.** Use distributive property to eliminate the parentheses; $\frac{3}{5}(15x + 10y)$. Remember, $\frac{3}{5} \bullet 15 = 9$ and $\frac{3}{5} \bullet 10 = 6$, resulting in the expression $9x + 6y$.

**13. d.** Use the distributive property to multiply each term inside the parentheses by $2x$; $(2x \bullet 7x) + (2x \bullet 5)$. Simplify by multiplying the coefficients of each term and adding the exponents of the like variables; $14x^2 + 10x$.

**14. e.** The formula for the area of a rectangle is *Area = length × width*. Since the length is $4y$ and the width is $5y + 8$, then multiply $4y(5y + 8)$. Use the distributive property to multiply $4y$ by both of the terms in the parentheses; $4y \bullet 5y + 4y \bullet 8$. Simplify by multiplying in each term; $20y^2 + 32y$.

**15. a.** Use order of operations and multiply first; $(3x)(4y) - (6x)(3y)$ becomes $12xy - 18xy$. Since both terms have exactly the same variables, they are *like terms* and can be combined. The final answer is $-6xy$.

**16. b.** Use the distributive property to multiply each term inside the parentheses by $-8p^3r$; $(-8p^3r \bullet 2p) - (-8p^3r \bullet 3r)$. Simplify by multiplying the coefficients of each term and adding the exponents of the like variables; $-16p^{3+1}r - (-24p^3r^{1+1})$. Change subtraction to addition and change the sign of the following term to its opposite; $-16p^4r + (+24p^3r^2)$. The final answer is $-16p^4r + 24p^3r^2$.

**17. c.** Use FOIL to multiply the binomials; $x \bullet x + 4 \bullet x + 2 \bullet x + 2 \bullet 4$. Simplify each term; $x^2 + 4x + 2x + 8$. Combine like terms; $x^2 + 6x + 8$.

**18. a.** Use FOIL to multiply the binomials; $2x \bullet x + 2x \bullet 8 - 1 \bullet x - 1 \bullet 8$. Simplify; $2x^2 + 16x - 1x - 8$. Combine like terms; $2x^2 + 15x - 8$.

**19. c.** The term *product* is a key word for multiplication. Use FOIL to multiply the binomials; $(2x \bullet 2x) + (2x \bullet 1) - (1 \bullet 2x) - (1 \bullet 1)$. Simplify each term; $4x^2 + 2x - 2x - 1$. Combine like terms; $4x^2 - 1$.

**20. e.** Since a binomial is being raised to the second power, write the binomial as two factors. Then multiply using the distributive property (FOIL); $(x - 1)^2 = (x - 1)(x - 1)$; $(x \bullet x) + (x \bullet 1) + (1 \bullet x) + (-1 \bullet -1)$. Simplify each term; $x^2 - x - x + 1$. Combine like terms; $x^2 - 2x + 1$.

**21. b.** Use FOIL to multiply the binomials; $(4x^2 \bullet 3x) + (8 \bullet 4x^2) + (2 \bullet 3x) + (2 \bullet -8)$. Simplify each term; $12x^3 - 32x^2 - 6x + 16$.

**22. d.** Use the distributive property to multiply the first term of the binomial, $x$, by each term of the trinomial, and then the second term of the binomial, $-3$, by each term of the trinomial; $(x \bullet x^2) + (x \bullet 4x) + (x \bullet 2) - (3 \bullet x^2) - (3 \bullet 4x) - (3 \bullet 2)$. Simplify by multiplying in each term; $x^3 + 4x^2 + 2x - 3x^2 - 12x - 6$. Use commutative property to arrange like terms next to each other; $x^3 + 4x^2 - 3x^2 + 2x - 12x - 6$. Combine like terms; $x^3 + x^2 - 10x - 6$.

**23. c.** Use the distributive property to multiply the first term of the binomial, $2x$, by each term of the trinomial, and then the second term of the binomial, $-1$, by each term of the trinomial; $(2x \bullet x^2) + (2x \bullet -3x) + (2x \bullet -4) - (1 \bullet x^2) - (1 \bullet -3x) - (1 \bullet -4)$. Simplify by

multiplying in each term; $2x^3 + -6x^2 + -8x - x^2 - (-3x) - (-4)$. Use commutative property to arrange like terms next to each other; $2x^3 + -6x^2 - x^2 + -8x + 3x + 4$. Combine like terms; $2x^3 - 7x^2 - 5x + 4$.

**24. e.** Use FOIL to multiply the first two binomials; $(x + 1)(x - 5)$; $(x \bullet x) + x(-5) + (1 \bullet x) + 1(-5)$. Simplify by multiplying in each term; $x^2 - 5x + 1x - 5$. Combine like terms; $x^2 - 4x - 5$. Multiply the third factor by this result; $(x + 4)(x^2 - 4x - 5)$. To do this, use the distributive property to multiply the first term of the binomial, $x$, by each term of the trinomial, and then the second term of the binomial, 4, by each term of the trinomial; $(x \bullet x^2) + (x \bullet -4x) + (x \bullet -5) + (4 \bullet x^2) + (4 \bullet -4x) + (4 \bullet -5)$. Simplify by multiplying in each term; $x^3 - 4x^2 - 5x + 4x^2 - 16x - 20$. Use commutative property to arrange like terms next to each other; $x^3 - 4x^2 + 4x^2 - 5x - 16x - 20$. Combine like terms; $x^3 - 21x - 20$.

**25. c.** The formula for the volume of a rectangular prism is *Volume = length × width × height*. Take each of the expressions for the dimensions and multiply them together. Use FOIL to multiply the first two binomials; $(x - 3)(x + 4)$; $(x \bullet x) + x(4) + (-3 \bullet x) + -3(4)$. Simplify by multiplying in each term; $x^2 + 4x + -3x - 12$. Combine like terms; $x^2 + x - 12$. Multiply the third factor by this result; $(2x + 1)(x^2 + x - 12)$. To do this, use the distributive property to multiply the first term of the binomial, $2x$, by each term of the trinomial, and then the second term of the binomial, 1, by each term of the trinomial; $(2x \bullet x^2) + (2x \bullet x) + (2x \bullet -12) + (1 \bullet x^2) + (1 \bullet x) + (1 \bullet -12)$. Simplify by multiplying in each term; $2x^3 + 2x^2 - 24x + x^2 + 1x - 12$. Use commutative property to arrange like terms next to each other; $2x^3 + 2x^2 + x^2 - 24x + 1x - 12$. Combine like terms. $2x^3 + 3x^2 - 23x - 12$.

# 8

# Factoring and Quadratic Equations

**F**actoring polynomials and solving quadratic equations are both critical and useful tools when working with algebra. Before starting this chapter, take the ten-question *Benchmark Quiz* to assess your understanding of both topics. The questions that appear here are similar to questions that you will find on important tests and will give you a good indication of where your strengths in these areas lie. When you are finished, check the answer key and use the answer explanations to help. Your Benchmark Quiz analysis will help you determine how much time you should spend practicing factoring and solving quadratic equations.

## BENCHMARK QUIZ

1. Factor the polynomial completely: $6x + 12$.
    a. $6(x + 1)$
    b. $6(x + 2)$
    c. $6(x + 6)$
    d. $6x(x + 2)$
    e. $6x(x + 6)$

2. Factor the polynomial completely: $4x^2 + 4x$.
   **a.** $4(x^2 + x)$
   **b.** $4x(x + 1)$
   **c.** $x(4x + 4)$
   **d.** $(4x + 1)(x + 4)$
   **e.** $(4x + 4)(x + 1)$

3. Factor the polynomial completely: $10b^3 + 5b - 15$.
   **a.** $2(5b^3 + 5b - 15)$
   **b.** $5(5b^3 + 5b - 10)$
   **c.** $5(2b^3 + b - 3)$
   **d.** $(5b^2 + 3)(2b - 5)$
   **e.** $(2b^2 + 3)(5b - 5)$

4. Factor the polynomial completely: $c^2 - 16$.
   **a.** $c(c - 4)$
   **b.** $c(c - 16)$
   **c.** $(c - 4)(c - 4)$
   **d.** $(c + 4)(c - 4)$
   **e.** $(c + 4)(c + 4)$

5. Factor the polynomial completely: $16x^2 - 25$.
   **a.** $16(x^2 - 25)$
   **b.** $4x(x - 5)$
   **c.** $(4x + 5)(4x - 5)$
   **d.** $(4x - 5)(4x - 5)$
   **e.** $(16x - 25)(16x + 25)$

6. Factor the polynomial completely: $x^2 + 8x + 7$.
   **a.** $x(x + 8)(x + 7)$
   **b.** $(x + 8)(x + 1)$
   **c.** $(x + 7)(x + 1)$
   **d.** $(x - 7)(x - 1)$
   **e.** $x^2(x + 7)$

7. Factor the polynomial completely: $x^2 + x - 20$.
   **a.** $x(x - 20)$
   **b.** $x(x - 5)$
   **c.** $(x - 20)(x + 1)$
   **d.** $(x - 5)(x - 4)$
   **e.** $(x - 4)(x + 5)$

8. Factor the polynomial completely: $2x^2 - 5x - 3$.
   a. $2x(x - 3)$
   b. $2x(x - 3)(x - 5)$
   c. $(2x - 1)(2x - 5)$
   d. $(2x - 3)(x + 1)$
   e. $(2x + 1)(x - 3)$

9. Solve for $x$: $x^2 + 2x - 24 = 0$.
   a. 4
   b. 12
   c. 4 or –6
   d. –4 or 6
   e. 2 or –12

10. Solve for $x$: $2x^2 + 3x = 4$.
    a. 3 or –4
    b. –1 or 4
    c. $\frac{3}{4} - \frac{\sqrt{41}}{4}$ or $\frac{3}{4} + \frac{\sqrt{41}}{4}$
    d. $-\frac{3}{4} - \frac{\sqrt{41}}{4}$ or $\frac{-3}{4} + \frac{\sqrt{41}}{4}$
    e. none of these

## BENCHMARK QUIZ SOLUTIONS

How did your memory of factoring and solving quadratic equations serve you? Check here to see if your answers match the correct solutions, and then plan out your route to factoring and quadratic equation success.

# ▶ Answers

**1. b.** Factor out the greatest common factor of 6 from each term. The binomial becomes $6(x + 2)$.

**2. b.** Factor out the greatest common factor of $4x$. The binomial becomes $4x(x + 1)$.

**3. c.** Factor out the greatest common factor of 5. The binomial becomes $5(2b^3 + b - 3)$.

**4. d.** This binomial is the difference between two perfect squares. The square root of $c^2$ is $c$ and the square root of 16 is 4. Therefore, the factors are $(c - 4)(c + 4)$.

**5. c.** This binomial is the difference between two perfect squares. The square root of $16x^2$ is $4x$ and the square root of 25 is 5. Therefore, the factors are $(4x + 5)(4x - 5)$.

**6. c.** Since all of the answer choices are binomials, factor the trinomial into two binomials. To do this, you will be doing a method that resembles FOIL backward (**F**irst terms of each binomial multiplied, **O**uter terms in each multiplied, **I**nner terms of each multiplied, and **L**ast term of each binomial multiplied.) **F**irst results in $x^2$, so the first terms must be:$(x\ )(x\ )$. **O**uter added to the **I**nner combines to $8x$, and the **L**ast is 7, so you need to find two numbers that add to $+8$ and multiply to $+7$. These two numbers would have to be $+1$ and $+7$. Therefore the factors are $(x + \underline{1})(x + \underline{7})$.

**7. e.** Since all of the answer choices are binomials, factor the trinomial into two binomials. To do this, you will be doing a method that resembles FOIL backward (**F**irst terms of each binomial multiplied, **O**uter terms in each multiplied, **I**nner terms of each multiplied, and **L**ast term of each binomial multiplied.) **F**irst results in $x^2$, so the first terms must be: $(x\ \_)(x\ \_)$. **O**uter added to the **I**nner combines to $1x$, and the **L**ast is $-20$, so you need to find two numbers that add to $+1$ and multiply to $-20$. These two numbers would have to be $-4$ and $+5$. Therefore the factors are $(x - \underline{4})(x + \underline{5})$.

**8. e.** Since all of the answer choices are binomials, factor the trinomial into two binomials. To do this, you will be doing a method that resembles FOIL backward (**F**irst terms of each binomial multiplied, **O**uter terms in each multiplied, **I**nner terms of each multi-plied, and **L**ast term of each binomial multiplied.) **F**irst results in $2x^2$, so the first terms must be: $(2x \_)(x \_)$. The **L**ast is $-3$, so you need to find two numbers that multiply to $-3$. The **O**uter added to the **I**nner combines to $-5x$, but in this case is a little different because of the $2x$ in the first binomial. When you multiply the Outer values, you are not just multiplying by $x$ but by $2x$. The only factors of $-3$ are $-1$ and $3$ or $1$ and $-3$, so try these values to see what placement of them works. The two numbers would then have to be $1$ and $-3$, placed as follows: $(2x + 1)(x - 3)$.

**9. c.** In order to solve a quadratic equation, make sure that the equation is in standard form. This equation is already in the correct form; $x^2 + 2x - 24 = 0$. Factor the left side of the equation; $(x - 4)(x + 6) = 0$. Set each factor equal to zero and solve for $x$; $x - 4 = 0$ or $x + 6 = 0$. The solutions to this equation are $x = 4$ or $x = -6$.

**10. d.** Put the equation in standard form; $2x^2 + 3x - 4 = 0$. Since this equation is not factorable, use the quadratic formula by identify-ing the value of $a$, $b$, and $c$ and then substituting into the formula. For this particular equation, $a = 2$, $b = 3$, and $c = -4$. The quadratic equation is $x = \frac{-b \pm \sqrt{b^2 - 4ac}}{2a}$. Substitute the values of $a$, $b$, and $c$ to get $x = \frac{-3 \pm \sqrt{3^2 - 4(2)(-4)}}{2(2)}$. This simplifies to $x = \frac{-3 \pm \sqrt{9 + 32}}{4}$ which becomes $x = \frac{-3 \pm \sqrt{41}}{4}$. So $x = \frac{-3}{4} \pm \frac{\sqrt{41}}{4}$. The solution is $\left\{\frac{-3}{4} - \frac{\sqrt{41}}{4}, \frac{-3}{4} + \frac{\sqrt{41}}{4}\right\}$.

## BENCHMARK QUIZ RESULTS

If you answered 8–10 questions correctly, you have pretty solid factoring skills and recall the steps to solving quadratic equations. Read through the lesson and focus on any areas you need to review. Then try the quiz at the end of the chapter to assess your complete comprehension of the content.

If you answered 4–7 questions correctly, you would certainly benefit by reviewing the concepts and procedures of factoring and solving quadratics. Take a look at the chapter, carefully reading through the skill building help, and note the sidebars that refer you to more in-depth practice, hints, and shortcuts. Work through the quiz at the end of the chapter to assess your progress.

If you answered 1–3 questions correctly, more time on factoring is the road to take. Begin with a careful read-through of the chapter, concentrating on the basics of factoring. You have probably seen some of the information prior to beginning using this book so take some time now to review and improve your knowledge. Take the quiz at the end of the chapter and use the answer explanations to clear up any confusion. You may also want to reference more comprehensive materials, such as LearningExpress's *Algebra Success in 20 Minutes a Day.*

## JUST IN TIME LESSON—FACTORING AND QUADRATIC EQUATIONS

This lesson will cover the basics of factoring and solving quadratic equations. The topics presented are

- factoring using the greatest common factor of the terms
- factoring the difference between two perfect squares
- factoring a trinomial in the form $ax^2 + bx + c$
- factoring a polynomial by grouping
- solving quadratic equations by factoring
- solving quadratic equations by the quadratic formula
- solving word problems involving quadratics

## FACTORS

The factors of a number or expression are the numbers and/or variables that can be multiplied to equal it. For example, the number 8 can be written as:

$1 \cdot 8 = 8$ or $2 \cdot 4 = 8$.
The numbers 1, 2, 4, and 8 are the factors of 8.

The expression $4x^2$ is the result of multiplying $4 \cdot x \cdot x$ so these are the factors of this expression.

## ▶ Greatest Common Factor

When factoring, you will often look for the greatest common factor of the terms in the expression. The greatest common factor, or GCF, is the greatest quantity that is common between the terms. Take a look at these examples.

> Find the GCF of 24 and 36.
> The factors of 24 are 1, 2, 3, 4, 6, 8, 12, and 24.
> The factors of 36 are 1, 2, 3, 4, 6, 9, 12, 18, and 36.
> Although the numbers 1, 2, 3, 4, 6, and 12 appear in both lists, the greatest of these is 12. Therefore, 12 is the GCF of 24 and 36.

SHORTCUT

**A quick way to find the GCF of two numbers is called *repeated division*. To do this, divide any common factor into the numbers, but work upside-down. Keep going until the only factor common to both numbers is 1. When done with the division, multiply together the numbers that divided both. For example:**

3 | 24  36    Divide both 24 and 36 by 3.

3 | 24  36    The result is 8 + 12.
     8  12

3 | 24  36    Divide 8 + 12 by 4.
4 | 8  12    The result is 2 and 3,
    2  3    which are only
          divisible by 1.

**Multiply 4 • 3 to get the greatest common factor.
Therefore, the GCF is 12.**

To find the GCF of expressions that involve variables, look for the common numerical factors and the number of common variables. For the terms $4x^2y$ and $6xy^2$, the common numerical factor is 2. For the variables, the first term has two $x$'s and one $y$. The second term has one $x$ and two $y$'s. Since the GCF only contains the number of factors that are common to both terms, the common variables are $xy$. Therefore, the GCF of $4x^2y$ and $6xy^2$ is $2xy$.

## FACTORING POLYNOMIALS

Factoring polynomials is the reverse of multiplying them together. The process of factoring polynomials usually falls into one of these categories: factoring out the GCF, the difference between two perfect squares, factoring a trinomial, and factoring by grouping. The goal of factoring is to always factor as much as you can. In other words, factor *completely*. Here are some examples of each type.

### ▶ *Factoring out the GCF*

In this type of problem, you will see that all the terms have something in common. This may be a numerical factor, variable factor, or a combination of both.

> *Examples:*
> 1. $2x^3 + 2 = 2(x^3 + 1)$
>    Put in front of the parentheses the common factor of 2. The terms remaining within parentheses are the original terms divided by 2. To check that the factoring is correct, multiply the factored answer using distributive property; $2(x^3 + 1) = 2 \bullet x^3 + 2 \bullet 1 = 2x^3 + 2$.
>    These factors are correct.
>
> 2. $8x^2y - 4xy + 2x = 2x(4xy - 2y + 1)$
>    Since the GCF is $2x$, factor out the GCF from each term by writing $2x$ on the outside of the parentheses and the remaining part of the terms inside the parentheses.

### ▶ *Factoring the Difference between Two Perfect Squares*

This type of situation will present two perfect square terms separated by subtraction. Numerical perfect squares are the result of multiplying two of the same integers together like $5 \bullet 5 = 25$ or $9 \bullet 9 = 81$. Variables that are perfect squares will have an exponent that is an even number. For example, since $x \bullet x = x^2$ then $\sqrt{x^2} = x$. In order to factor the difference between two squares, take the square root of each perfect square and express the difference between them as one factor and their sum as the other.

> *Examples:*
> 1. $x^2 - 9 = (x - 3)(x + 3)$ because $\sqrt{x^2} = x$ and $\sqrt{9} = 3$.
>    To check to make sure these factors are correct, multiply using FOIL (distributive property).

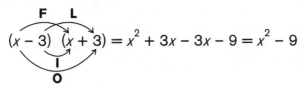

$$(x - 3)\ (x + 3) = x^2 + 3x - 3x - 9 = x^2 - 9$$

These factors are correct.

**2.** $4c^2 - 49 = (2c - 7)(2c + 7)$ because $\sqrt{4c^2} = 2c$ and $\sqrt{49} = 7$.

 RULE BOOK
The difference of two squares is factored $x^2 - y^2 = (x - y)(x + y)$. The sum of two squares is *not* factorable.

## ▶ *Factor the Trinomial in the Form $ax^2 + bx + c$*

In order to factor a binomial in this form, use backward FOIL. For situations where the value of $a = 1$, the factors will contain two numbers whose sum is $b$ and whose product is $c$. Take the following examples.

*Examples:*
**1.** $x^2 + 5x + 6$

Since $x^2 = x \bullet x$, the leading terms in the parentheses will both be $x$'s. In this problem $a = 1$, so find two numbers that add to $b$, which is 5, and multiply to equal $c$, which in this case is 6. These two numbers are 2 and 3. This makes the factors $(x + 2)(x + 3)$. To check to make sure these factors are correct, multiply them using FOIL (distributive property).

$$(x - 2)\ (x + 3) = x^2 + 3x - 2x + 6 = x^2 + 5x + 6$$

These factors are correct.

**2.** $x^2 - 3x - 10$

In this case you are looking for two numbers whose sum is $-3$ and whose product is $-10$. These two numbers are 2 and $-5$. Therefore the factors are $(x + 2)(x - 5)$.

To factor a quadratic equation in the form $ax^2 + bx + c = 0$ (the value of $a = 1$), the factors will be in the form $(x + \_)(x + \_)$ where the numerical factors have a **sum** of $b$ and a **product** of $c$.

**3.** $x^2 - 9x + 18$

This trinomial has a $b$ value of $-9$ and a $c$ value of 18. Two numbers that add to $-9$ and whose product is 18 are $-3$ and $-6$. The factors are $(x - 3)(x - 6)$.

When factoring a trinomial in this form where the value of $a$ is not 1, it is sometimes a little more work. Very often, using trial and error with FOIL can lead you to the correct factors. Here is an example.

**4.** $2x^2 + 5x - 3$

Factor using FOIL backward, but keep in mind that since $a = 2$ the **O**utside terms when multiplied will be affected. First find factors of $2x^2$. These will be $2x$ and $x$ and will be the leading terms in the parentheses. Now you need factors of $-3$ that will combine with $2x$ and $x$ to make a middle term of $5x$. Find the factors by trial and error.

Begin by trying $-3$ and 1 as the factors of $-3$. The factors would be $(2x - 3)(x + 1)$. When multiplied together this equals $2x^2 + 2x - 3x - 3$. Combine like terms to get $2x^2 - 1x - 3$, which is equal to $2x^2 - x - 3$. Since this is not the original trinomial, these are not the correct factors.

Switch the $-3$ and 1 to the opposite binomials. The factors would be $(2x + 1)(x - 3)$. When multiplied together this equals $2x^2 - 6x + 1x - 3$. Combine like terms to get $2x^2 - 5x - 3$. Since this is not the original trinomial, these are not the correct factors.

Switch the signs on the numbers to make the factors of $-3$ to be 3 and $-1$. The factors of the trinomial would then be $(2x + 3)(x - 1)$. When multiplied together this equals $2x^2 - 2x + 3x - 3$. Combine like terms to get $2x^2 + 1x - 3$, which is equal to $2x^2 + x - 3$. Since this is not the original trinomial, these are not the correct factors.

Switch the $-1$ and 3 to the opposite binomials. The factors become $(2x - 1)(x + 3)$ When multiplied together this equals $2x^2 + 6x - 1x - 3$. Combine like terms. $2x^2 + 5x - 3$. Since this is the original trinomial, these are the correct factors.

### ▶ Factoring By Grouping

Sometimes you may encounter a polynomial with four terms. The strategies discussed previously may not work with this situation. Another factoring technique to try in this case is factoring by grouping. Here you will pair together terms with a common factor. After factoring out that common factor from each pair, the terms left in the parentheses are often the same. These terms can then be factored out to create the factored polynomial. Note the following example of factoring by grouping.

Take the polynomial $ax + bx - 4a - 4b$. The first pair of terms have a common factor of $x$ and the second pair of terms have a common factor of $-4$. Factor each pair separately. The resulting expression is $x(a + b) - 4(a + b)$. Now, each of these terms has a factor of $(a + b)$. Factoring $a + b$ out in front results in the expression $(a + b)(x - 4)$ which gives the factors of the polynomial.

SHORTCUT

**Factoring by grouping is usually the most helpful in expressions that have more than three terms.**

## ▶ Factoring Completely

Many times you will be presented with a polynomial where the factoring takes place over more than one step. It is very important that polynomial expressions are factored completely and that you go as far as you can go when working with this type of situation. Here are some examples where the factoring takes place over more than one step.

*Examples:*
**1.** $2x^2 - 128$

First factor a 2 out of both terms. The expression then becomes $2(x^2 - 64)$. Located in parentheses is the difference between two squares, which is also factorable. The final factors are $2(x - 8)(x + 8)$.

RULE BOOK

**When factoring polynomials, use the following steps:**

1. **Always look for a greatest common factor of each term first.**
2. **Look for the difference between two perfect squares.**
3. **Factor the trinomial using FOIL backward.**
4. **If there are four or more terms, try factoring by grouping.**

**2.** $d^4 - 16$

Since these terms do not have a common factor, then start by factoring the difference between two squares. $(d^2 - 4)(d^2 + 4)$. Of these two factors, the first binomial is also the difference between two squares and becomes $(d - 2)(d + 2)$. The second binomial is the sum of two squares and is not factorable. Therefore, the factors are $(d - 2)(d + 2)(d^2 + 4)$.

**3.** $2n^3 - 14n^2 + 24n$

First, factor out the common factor of $2n$ from each term. This leaves you with $2n(n^2 - 7n + 12)$. The trinomial in the parentheses can be factored. Two numbers that add to $-7$ and whose product is 12 are $-3$ and $-4$, so the factors are $(x - 3)(x - 4)$. Therefore the factors for the entire polynomial are $2n(x - 3)(x - 4)$.

 EXTRA HELP

The comprehensive resource LearningExpress's *Algebra in 20 Minutes a Day* contains even more on this topic. Take a look at Lesson 15: Factoring Polynomials and Lesson 16: Using Factoring for additional help and clarity.

## ▌ *Solving Quadratic Equations*

An equation in the form $0 = ax^2 + bx + c$, where $a$, $b$, and $c$ are real numbers, is a quadratic equation. Quadratic equations contain a term in which the exponent on $x$ is two, in other words, it contains an $x^2$. Two ways to solve quadratic equations algebraically are factoring, if it is possible for that equation, or by using the quadratic formula.

 GLOSSARY

**ZERO PRODUCT PROPERTY** If an equation is set equal to zero, then at least one of the factors must equal zero. If $xy = 0$, then either $x = 0$ or $y = 0$.

**By Factoring.** In order to factor the quadratic equation, it first needs to be in standard form. This form is $0 = ax^2 + bx + c$. In most cases, the factors of the equations involve two numbers whose sum is $b$ and product is $c$. Be sure to factor the quadratic completely. When factored, set each factor equal to zero and solve for $x$. In this step you are using the Zero Product Property as mentioned above. The resulting values are called the *roots* of the equation. Here are a few examples of how to deal with a factorable quadratic equation.

*Examples:*
**1.** Solve the following for $x$:
$x^2 - 100 = 0$
This equation is already in standard form. This equation is a special

case; it is the difference between two perfect squares. To factor this, find the square root of both terms.

The square root of the first term, $x^2$ is $x$.

The square root of the second term 100 is 10.

Then two factors are $x - 10$ and $x + 10$.

The equation $x^2 - 100 = 0$

then becomes $(x - 10)(x + 10) = 0$.

Set each factor equal to zero and solve:

$x - 10 = 0$ or $x + 10 = 0$.

$x = 10$ or $x = -10$

The solution is $\{10, -10\}$.

2. $x^2 + 14x = -49$

This equation needs to be put into standard form by adding 49 to both sides of the equation: $x^2 + 14x + 49 = -49 + 49$

$$x^2 + 14x + 49 = 0.$$

The factors of this trinomial will be two numbers whose sum is 14 and whose product is 49. The factors are $x + 7$ and $x + 7$ because 7 + 7 = 14 and $7 \bullet 7 = 49$. The equation becomes $(x + 7)(x + 7) = 0$. Set each factor equal to zero and solve $x + 7 = 0$ or $x + 7 = 0$; $x = -7$ or $x = -7$.

Because both factors were the same, this was a perfect square trinomial. The solution is $\{-7\}$.

3. $x^2 = 42 + x$

This equation needs to be put into standard form by subtracting 42 and $x$ from both sides of the equation:

$$x^2 - x - 42 = 42 - 42 + x - x$$
$$x^2 - x - 42 = 0.$$

Since the sum of 6 and $-7$ is $-1$, and their product is $-42$, the equation factors to $(x + 6)(x - 7) = 0$.

Set each factor equal to zero and solve: $x + 6 = 0$ or $x - 7 = 0$; $x = -6$ or $x = 7$.

The solution is $\{-6, 7\}$.

**By Quadratic Formula.** If the equation is not factorable, or if you are having trouble finding the factors, use the quadratic formula. Solving by using the quadratic formula will work for any quadratic equation and is necessary for those that are not factorable.

The quadratic formula is $x = \frac{-b \pm \sqrt{b^2 - 4ac}}{2a}$.

To use this formula, first put the equation into standard form. Then identify $a$, $b$, and $c$ from the equation and substitute these values into the quadratic formula.

*Example:*

Solve for $x$: $x^2 + 4x = 3$.

Put the equation in standard form: $x^2 + 4x - 3 = 0$.

Since this equation is not factorable, use the quadratic formula by identifying the values of $a$, $b$, and $c$ and then substituting into the formula. For this particular equation, $a = 1$, $b = 4$, and $c = -3$.

$$x = \frac{-b \pm \sqrt{b^2 - 4ac}}{2a}$$

$$x = \frac{-4 \pm \sqrt{4^2 - 4(1)(-3)}}{2(1)}$$

$$x = \frac{-4 \pm \sqrt{16 + 12}}{2}$$

$$x = \frac{-4 \pm \sqrt{28}}{2}$$

$$x = \frac{-4}{2} \pm \frac{2\sqrt{7}}{2} \quad \text{(Remember that } \sqrt{28} \text{ reduces to } 2\sqrt{7}.\text{)}$$

$$x = -2 \pm \sqrt{7}$$

The solution is $\{-2 + \sqrt{7}, -2 - \sqrt{7}\}$.

These roots are irrational because they contain a radical in the simplified solution. The square root of any non-perfect square is irrational.

 RULE BOOK

If a quadratic equation is difficult to factor or is unfactorable, use the quadratic formula to find the roots; $x = \frac{-b \pm \sqrt{b^2 - 4ac}}{2a}$.

 EXTRA HELP

For more help solving quadratic equations, visit the website www.cool-math.com and click on **Ages 13–100**. Under **Calculators** select **Quadratic** to get a calculator that will solve for the real solutions to a quadratic by plugging in the values of $a$, $b$, and $c$ from the equation.

## QUADRATIC EQUATIONS AND WORD PROBLEMS

The following is an example of a word problem incorporating quadratic equations.

A rectangular garden is surrounded by a walkway that has the same width all the way around. The total area of the walkway only is 500 square feet. The dimensions of the garden are 16 feet by 24 feet. How many feet is the width of the walkway?

Start by drawing a picture of the garden and walkway. It may look similar to the following diagram.

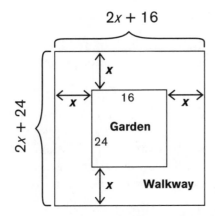

Let $x$ = the width of the walkway. Since the width of the garden is 16, the width of the garden and the walkway is $x + x + 16$ or $2x + 16$. Since the length of the garden is 24, the length of the garden and the walkway together is $x + x + 24$ or $2x + 24$. The total area for the garden and the walkway together is 884 square feet, 500 square feet added to $16 \times 24 = 384$ square feet for the garden. Area of a rectangle is *length times width*, so multiply the expressions together and set them equal to the total area of 884 square feet. $(2x + 16)(2x + 24) = 884$.

Multiply the binomials using the distributive property (FOIL):
$$4x^2 + 32x + 48x + 384 = 884.$$
Combine like terms: $\quad 4x^2 + 80x + 384 = 884$.
Subtract 884 from both sides:
$$4x^2 + 80x + 384 - 884 = 884 - 884.$$
Simplify: $\quad 4x^2 + 80x - 500 = 0$.
Factor the trinomial completely:
$$4(x^2 + 20x - 125) = 0$$
$$4(x - 5)(x + 25) = 0.$$
Set each factor equal to zero and solve:
$$4 \neq 0 \text{ or } x - 5 = 0 \text{ or } x + 25 = 0;$$
$$x = 5 \text{ or } x = -25.$$
Reject the negative solution because you will not have a negative length.
The width is 5 feet.

 SHORTCUT

When solving an equation by factoring, any number alone that is a factor will not give a solution to the equation. Note in the above example, when the factors were set equal to zero, the number 4 did not yield an answer; $4 \neq 0$ so this is not part of the solution set.

## TIPS AND STRATEGIES

Use the following hints and highlights to help you progress through the unit. Take special note of factoring shortcuts to help simplify and solve quadratic expressions and equations.

- When factoring polynomials, first look for a common factor between terms.
- When factoring a trinomial in the form $ax^2 + bx + c$ when $a = 1$, the sum of the factors is $b$ and the product of the terms is $c$.
- Know how to factor the difference between two squares: $x^2 - y^2 = (x + y)(x - y)$.
- When factoring a perfect square trinomial with all positive terms, remember that $x^2 + 2xy + y^2 = (x + y)(x + y) = (x + y)^2$.
- When factoring a perfect square trinomial with a negative middle term, remember that $x^2 - 2xy + y^2 = (x - y)(x - y) = (x - y)^2$.
- To solve a quadratic equation, first put the equation in the form $ax^2 + bx + c = 0$.
- The solution(s) to a quadratic equation are called the *roots* of the equation. These are values of $x$ where $y = 0$ for that equation.
- When solving a quadratic equation, first try to factor. If the equation is difficult to factor or is not factorable, use the quadratic formula to find the roots.
- When solving word problems involving quadratic equations, always start by defining the variable $x$ and use the same techniques as non-word problems (factoring or quadratic formula).
- Many factoring problems can be solved by trial and error, so take advantage by trying the possible answer choices on multiple-choice tests.
- When studying algebra you need to be an active learner; the more practice, the better.

## EXTRA HELP

**For many more examples and extra practice check out Lesson 17: Solving Quadratic Equations in LearningExpress's** *Algebra Success in 20 Minutes a Day.*

## CHAPTER QUIZ

The chapter quiz will give you more practice factoring and solving quadratic equations. See how you do by checking your answers with the explanations when you have completed the quiz.

1. What is the GCF of $3x^2$ and $9x^2y$?
   a. 3
   b. $3x$
   c. $3x^2$
   d. $9x$
   e. $9x^2y$

2. Factor completely $12ab^2c - 16a^2bc - 4abc$.
   a. $2a(6b^2c - 8abc - 2bc)$
   b. $4a(3b^2c - 4abc - bc)$
   c. $4ab(3bc - 4ac - c)$
   d. $4abc(3b - 4a)$
   e. $4abc(3b - 4a - 1)$

3. Factor the binomial: $x^2 - 25$.
   a. $(x + 5)(x + 5)$
   b. $(x - 5)(x - 5)$
   c. $(x + 5)(x - 5)$
   d. $x(x - 25)$
   e. $(x - 10)(x + 15)$

4. If $x - 1$ is one factor of the trinomial $3x^2 - 5x + 2$, what is the other factor?
   a. $x - 5$
   b. $x - 2$
   c. $3x - 1$
   d. $3x - 2$
   e. $3x + 2$

5. Which of the following are factors of the binomial: $x^2 + 16$?
   a. $(x + 4)(x + 4)$
   b. $(x - 4)(x - 4)$
   c. $(x + 8)(x + 2)$
   d. $(x - 8)(x - 2)$
   e. none of these

**6.** Factor the binomial: $2x^2 - 10xz$.
  **a.** $2(x^2 - 5xz)$.
  **b.** $2x(x - 5z)$
  **c.** $(2x - 1)(x - 10z)$
  **d.** $(2x - 5z)(x + 2)$
  **e.** $2z(x - 2)(x + 5)$

**7.** Factor the trinomial: $3y^3 + 6xy^2 + 9y$.
  **a.** $y(3y^2 + 2y + 9)$
  **b.** $3y^2(y + 2x + 3)$
  **c.** $3y(y^2 + 2xy + 3)$
  **d.** $3y^3(1 + 2x + 3)$
  **e.** $3y(y^3 + 2xy + 3)$

**8.** Factor the trinomial: $x^2 + 4x + 4$.
  **a.** $(x + 1)(x + 4)$
  **b.** $(x + 2)(x + 2)$
  **c.** $(x - 1)(x + 4)$
  **d.** $(x - 4)(x - 1)$
  **e.** $(x - 2)(x - 2)$

**9.** Factor the trinomial: $x^2 + 3x + 2$.
  **a.** $(x + 1)(x + 2)$
  **b.** $(x - 1)(x - 2)$
  **c.** $(x - 2)(x + 3)$
  **d.** $(x + 2)(x + 3)$
  **e.** $(x + 1)(x + 3)$

**10.** Factor the trinomial: $x^2 + x - 30$.
  **a.** $(x + 6)(x - 5)$
  **b.** $(x - 3)(x - 10)$
  **c.** $(x - 6)(x + 5)$
  **d.** $(x - 15)(x + 2)$
  **e.** $(x + 10)(x - 3)$

**11.** Factor the trinomial: $3x^2 - 10x + 3$.
  **a.** $(3x - 1)(x + 3)$
  **b.** $(3x - 1)(x - 3)$
  **c.** $(3x - 3)(x - 1)$
  **d.** $(3x + 3)(x + 1)$
  **e.** $(3x + 1)(x - 3)$

**12.** Factor the binomial completely: $2x^2 - 98$.
a. $2(x^2 - 49)$
b. $(2x - 7)(x + 7)$
c. $2(x - 7)(x + 7)$
d. $2x(x - 7)(x + 7)$
e. $2(x - 7)(x + 7)$

**13.** Factor the polynomial completely: $4x^3 - 4ax^2 + 8x^2 - 8ax$.
a. $(4x^2 + 8x)(x - a)$
b. $(4ax^2 - 2)(x - a)$
c. $4x(x + 2)(x - a)$
d. $4x(x - 2)(x + a)$
e. $(4ax - a)(8x - x)$

**14.** The area of Mr. Murphy's rectangular family room is represented by the expression $x^2 - 121$. Which of the following binomials could represent the length and the width of the room?
a. $(x + 11)$ and $(x + 11)$
b. $(x - 11)$ and $(x - 11)$
c. $(x + 11)$ and $(x - 11)$
d. $x$ and $(x - 121)$
e. $(x - 100)$ and $(x - 21)$

**15.** The square root of $x^2 + 18x + 81$ is
a. $x - 3$.
b. $x + 3$.
c. $x - 9$.
d. $x + 9$.
e. $x^2 + 9$.

**16.** Factor the polynomial completely: $yz - 7m + ym - 7z$.
a. $(y - 7)(z + m)$
b. $(y + 7)(z - m)$
c. $(y + 7)(z + m)$
d. $(y - 7)(z - m)$
e. $y(y + 7)(z - m)$

**17.** The area of a parallelogram can be expressed as the binomial $3x^2 - 18x$. Which could be the length of the base and the height of the parallelogram?
   **a.** $3x$ and $(x^2 - 6x)$
   **b.** $3x$ and $(x - 6)$
   **c.** $(3x - 1)$ and $(x - 18)$
   **d.** $(3x - 6)$ and $(x + 3)$
   **e.** $(3x + 6)$ and $(x + 3)$

**18.** Which of the following is the solution set of the equation $x(x - 5) = 0$?
   **a.** $\{0\}$
   **b.** $\{5\}$
   **c.** $\{0, -5\}$
   **d.** $\{0, 5\}$
   **e.** $\{-5\}$

**19.** What is the solution set of the equation $(x - 2)(2x - 5) = 0$?
   **a.** $\{-2, -5\}$
   **b.** $\{2, 5\}$
   **c.** $\{2, \frac{2}{5}\}$
   **d.** $\{2, \frac{5}{2}\}$
   **e.** $\{-2, -\frac{5}{2}\}$

**20.** What are the roots of the equation $x^2 = 196$?
   **a.** 14
   **b.** 14 or −14
   **c.** −14
   **d.** 4 or 49
   **e.** 98 or − 98

**21.** What is the positive root of the equation $x^2 + 4x = 21$?
   **a.** 3
   **b.** 4
   **c.** 7
   **d.** 21
   **e.** 84

**22.** What are the roots of the equation $x^2 - 8x = -5$?
   **a.** −8 or 5
   **b.** 8 or −5
   **c.** $8 - 2\sqrt{11}$ or $8 + 2\sqrt{11}$
   **d.** $4 - \sqrt{44}$ or $4 + \sqrt{44}$
   **e.** $4 - \sqrt{11}$ or $4 + \sqrt{11}$

23. Peter has a rectangular yard whose length is 5 m more than the width. If the area of his yard is 500 m², what is the length of his yard?
    a. 5
    b. 10
    c. 20
    d. 25
    e. 50

24. The square of a number added to 64 equals 16 times the number. What is the number?
    a. –4
    b. 8
    c. –8
    d. 4
    e. 32

25. A rectangular pool has a width of 25 feet and a length of 40 feet. A deck with a uniform width surrounds it. If the area of the deck and the pool together is 1,584 square feet, what is the width of the deck?
    a. 2 ft
    b. 2.5 ft
    c. 3 ft
    d. 4 ft
    e. 5 ft

## ANSWERS

Check your progress through factoring and quadratic equations with the following explanations. For even more extra help, reference the more extensive materials mentioned in the text.

1. c. The GCF is the greatest common factor of both terms. Each term has a numerical factor of 3 and a variable factor of $x^2$. Therefore the GCF is $3x^2$.

2. e. In order to factor this trinomial completely, first look for the greatest common factor, or GCF, of the three terms. Each term contains a numerical factor of 4 and variable factors of $abc$. Therefore the GCF is $4abc$. If you divide each term by $4abc$ you are left with $3b - 4a - 1$ in the parentheses. Therefore the factored form is $4abc(3b - 4a - 1)$.

**3. c.** Since this binomial is the difference between two perfect squares, the square root of $x^2$ is $x$ and the square root of 25 is 5. The factors expressed as the sum and the difference of these square roots are $(x + 5)$ and $(x - 5)$.

**4. d.** In this case, you know one of the factors of the trinomial. To find the other factor, first use the fact that the first term in the trinomial is $3x^2$; therefore the answer choices are narrowed down to **c**, **d**, or **e**, ($x$ times $3x$ equals $3x^2$.) Remember that the product of the last terms in each of the binomials is equal to $c$ in the equation. Since $c$ is equal to 2 and the known factor is $x - 1$, then $-1$ would be multiplied by $-2$ to get a value of $+2$. This leads you to choice **d**, which is $3x - 2$, as the solution. Use FOIL (distributive property) to check your answer; $(x - 1)(3x - 2) = 3x^2 - 2x - 3x + 2 = 3x^2 - 5x + 2$. Choice **d** is correct.

**5. e.** This binomial is not factorable. It is the *sum* of two perfect squares. Do not confuse it with $x^2 - 16$ which is factorable because it is the *difference* between two perfect squares. If you multiply out any of the choices, none of them will equal $x^2 + 16$.

**6. b.** First look for factors that both terms have in common. $2x^2$ and $10xz$ both have factors of 2 and $x$. Factor out the greatest common factor, $2x$, from each term. $2x^2 - 10x = 2x(x - 5z)$. To check an answer like this, multiply through using the distributive property; $2x(x - 5z) = (2x \bullet x) - (2x \bullet 5z) = 2x^2 - 10z$. This question checked because the result is the same as the original binomial.

**7. c.** In order to factor this trinomial completely, first look for the greatest common factor, or GCF, of the three terms. Each term contains a numerical factor of 3 and variable factor of $y$. Therefore the GCF is $3y$. If you divide each term by $3y$ you are left with $y^2 + 2xy + 3$ in the parentheses. Therefore the factored form is $3y(y^2 + 2xy + 3)$.

**8. b.** Since each of the answer choices are binomials, factor the trinomial into two binomials. To do this, you will be doing a method that resembles FOIL backward (**F**irst terms of each binomial multiplied, **O**uter terms in each multiplied, **I**nner terms of each multiplied, and **L**ast term of each binomial multiplied.) **F**irst results in $x^2$, so the first terms must be:$(x \_)(x \_)$. **O**uter added to the **I**nner combines to $4x$, and the **L**ast is 4, so you need to find two numbers that add to $+4$ and multiply to $+4$. These two numbers would have to be +2 and +2. Therefore the factors are $(x + 2)(x + 2)$. Since the factors of the trinomial are the same, this is an example of a perfect square trinomial.

**9. a.** First check to see if there is a common factor in each of the terms or if it is the difference between two perfect squares, and it is neither of these. The next step would be to factor the trinomial into two binomials. To do this, you will be doing a method that resembles FOIL backward. (**F**irst terms of each binomial multiplied, **O**uter terms in each multiplied, **I**nner terms of each multiplied, and **L**ast term of each binomial multiplied.) **F**irst results in $x^2$, so the first terms must be $(x \_)(x \_)$. **O**uter added to the **I**nner combines to $3x$, and the **L**ast is 2, so you need to find two numbers that add to +3 and multiply to +2. These two numbers would have to be +1 and +2. Therefore the factors are $(x + 1)(x + 2)$.

**10. a.** Factor the trinomial into two binomials. To do this, you will be doing a method that resembles FOIL backward (**F**irst terms of each binomial multiplied, **O**uter terms in each multiplied, **I**nner terms of each multiplied, and **L**ast term of each binomial multiplied.) **F**irst results in $x^2$, so the first terms must be $(x \_)(x \_)$. **O**uter added to the **I**nner combines to $1x$, and the **L**ast is –30, so you need to find two numbers that add to +1 and multiply to –30. These two numbers would have to be –5 and +6. Thus, the factors are $(x + 6)$ and $(x - 5)$.

**11. b.** First check to see if there is a common factor in each of the terms or if it is the difference between two perfect squares, and it is neither of these. The next step would be to factor the trinomial into two binomials. To do this, you will be doing a method that resembles FOIL backward. (**F**irst terms of each binomial multiplied, **O**uter terms in each multiplied, **I**nner terms of each multiplied, and **L**ast term of each binomial multiplied.) **F**irst results in $3x^2$, so the first terms must be $(3x \_)(x \_)$. The **L**ast is +3, so you need to find two numbers that multiply to +3. The **O**uter added to the **I**nner combines to $-10x$ but in this case is a little different because of the $3x$ in the first binomial. When you multiply the Outer values, you are not just multiplying by $x$ but by $3x$. The two numbers would then have to be –1 and –3, placed as follows: $(3x - 1)(x - 3)$.

**12. c.** In order to factor completely, first look for a common factor between the two terms. Since 2 can be divided into both terms, factor 2 out of the binomial; $2(x^2 - 49)$. The expression in the parentheses is the difference between two perfect squares. Since the square root of $x^2$ is $x$ and the square root of 49 is 7, the factors are $(x - 7)(x + 7)$. The final answer factored completely is $2(x - 7)(x + 7)$.

**13. c.** Since there are more than 3 terms, try factoring by grouping. The first pair of terms each has a common factor of $4x^2$, so factor $4x^2$ out of those terms. The second pair of terms each has a factor of $8x$, so factor $8x$ out of those terms. At this point the expression should look like $4x^2(x - a) + 8x(x - a)$. Each of these terms now has a common factor of $(x - a)$. Factoring again results in the binomials $(x - a)(4x^2 + 8x)$. Since the second binomial can still be factored to $4x(x + 2)$, the final answer is $4x(x - a)(x + 2)$ which is equivalent to choice **c.**

**14. c.** Since the formula for area is *Area = length times width*, factor the binomial to find the dimensions of the family room. Since this binomial is the difference between two perfect squares, the factors are $(x + 11)$ and $(x - 11)$, because $\sqrt{(x^2)} = x$ and $\sqrt{121} = 11$.

**15. d.** Since you are looking for the square root, you are trying to find the expression that when multiplied by itself equals this trinomial. Look for the factors of the trinomial $x^2 + 18x + 81$ by trying to find two numbers whose product is 81 and whose sum is 18. These numbers are 9 and 9. Therefore the factors of the trinomial are $(x + 9)(x + 9)$. This is a perfect square trinomial and the square root is $x + 9$.

**16. a.** Since there are more than 3 terms, try factoring by grouping. Rewrite the terms in the order $yz + ym - 7m - 7z$ so that the consecutive terms have a factor in common. The first pair of terms each has a common factor of $y$, so factor $y$ out of those terms. The second pair of terms each has a factor of $-7$, so factor $-7$ out of those terms. Keep in mind that factoring out $-7$ will change the signs of the terms you are dividing it out of. At this point the expression should look like $y(z + m) - 7(m + z)$. Remember that $z + m = m + z$ so each of these terms now has a common factor of $(m + z)$. Factoring again results in the binomials $(m + z)(y - 7)$ which is equivalent to choice **a.**

**17. b.** To find the base and the height of the parallelogram, find the factors of this binomial. First look for factors that both terms have in common; $3x^2$ and $18x$ both have a factor of 3 and $x$. Factor out the greatest common factor, $3x$, from each term; $3x^2 - 18x = 3x(x - 6)$.

**18. d.** Since the equation is in already in factored form and set equal to zero, set each of the factors individually equal to zero and solve for $x$; $x = 0$ or $(x - 5) = 0$. This gives a solution of 0 or 5, which is choice **d.**

**19. d.** Since the equation is already in factored form and set equal to zero, set each of the factors individually equal to zero and solve for $x$; $(x - 2) = 0$ or $(2x - 5) = 0$. The first equation gives a solution of 2. Take the second equation and add 5 to both sides of the equal sign; $2x - 5 + 5 = 0 + 5$. This simplifies to $2x = 5$. Divide both sides of the equal sign by 2; $\frac{2x}{2} = \frac{5}{2}$. Therefore this solution is $x = \frac{5}{2}$. The solution set is $\{2, \frac{5}{2}\}$.

**20. b.** Put the equation in standard form by subtracting 196 from both sides of the equal sign to get $x^2 - 196 = 0$. The left side of the equation is the difference between two squares. Since the square root of $x^2$ is $x$ and the square root of 196 is 14, the equation becomes $(x - 14)(x + 14) = 0$. Set each factor equal to zero and solve for $x$; $x - 14 = 0$ or $x + 14 = 0$. The solutions are 14 or –14.

**21. a.** Put the equation in standard form by subtracting 21 from both sides of the equal sign; $x^2 + 4x - 21 = 0$. Factor the trinomial on the left side of the equal sign into two binomials using FOIL. Two numbers that add to 4 and multiply to –21 are –3 and 7. Therefore, the equation factors to $(x - 3)(x + 7) = 0$. Set each factor equal to zero and solve for $x$; $x - 3 = 0$ or $x + 7 = 0$. The solutions are 3 or –7. Since you are looking for the positive root, the answer is 3.

**22. e.** Put the equation in standard form; $x^2 - 8x + 5 = 0$. Since this equation is not factorable, use the quadratic formula by identifying the value of $a$, $b$, and $c$ and then substituting into the formula. For this particular equation, $a = 1$, $b = -8$, and $c = 5$. The quadratic equation is $x = \frac{-b \pm \sqrt{b^2 - 4ac}}{2a}$. Substitute the values of $a$, $b$, and $c$ to get $x = \frac{-(-8) \pm \sqrt{(-8)^2 - 4(1)(5)}}{2(1)}$. This simplifies to $x = \frac{8 \pm \sqrt{64 - 20}}{2}$ which becomes $x = \frac{8 \pm \sqrt{44}}{2}$. So $x = \frac{8}{2} \pm \frac{2\sqrt{11}}{2}$. Reducing gives the solution is $\{4 - \sqrt{11}, 4 + \sqrt{11}\}$.

**23. d.** Let $x$ = the width of his yard. Therefore, $x + 5$ = the length of the yard. Since the area is 500 m², and area is *length times width*, the equation is $x(x + 5) = 500$. Use the distributive property to multiply the left side; $x^2 + 5x = 500$. Subtract 500 from both sides; $x^2 + 5x - 500 = 500 - 500$. Simplify; $x^2 + 5x - 500 = 0$. Factor the result; $(x - 20)(x + 25) = 0$. Set each factor equal to 0 and solve; $x - 20 = 0$ or $x + 25 = 0$. The solutions are $x = 20$ or $x = -25$. Reject the solution of $-25$ because a distance will not be negative. Since the width is 20 m, the length is 25 m.

**24. b.** Let $x$ = the number. The statement "The square of a number added to 64 equals 16 times the number" translates to the equation $x^2 + 64 = 16x$. Put the equation in standard form and set it equal to zero; $x^2 - 16x + 64 = 0$. Factor the left side of the equation; $(x - 8)(x - 8) = 0$. Set each factor equal to zero and solve; $x - 8 = 0$ or $x - 8 = 0$. The solutions are $x = 8$ or $x = 8$, so the number is 8.

**25. d.** Let $x$ = the width of the deck. Therefore, $x + x + 25$ or $2x + 25$ is the width of the entire figure. In the same way, $x + x + 40$ or $2x + 40$ is the length of the entire figure. The area of a rectangle is *length × width*, so use $A = l \times w$. Substitute into the equation; $1,584 = (2x + 40)(2x + 25)$

Multiply using FOIL: $1,584 = 4x^2 + 50x + 80x + 1,000$.
Combine like terms:   $1,584 = 4x^2 + 130x + 1,000$.
Subtract 1,584 from both sides:
$$1,584 - 1,584 = 4x^2 + 130x + 1000 - 1,584;$$
$$0 = 4x^2 + 130x - 584.$$
Factor out the GCF of 2:
$$0 = 2(2x^2 + 65x - 292).$$
Factor the remaining trinomial:
$$0 = 2(2x + 73)(x - 4).$$
Set each factor equal to zero and solve:
$$2 \neq 0 \text{ or } 2x + 73 = 0 \text{ or } x - 4 = 0$$
$$2x = -73 \qquad\qquad x = 4$$
$$x = -36.5$$
Since we are solving for a length, the solution of $-36.5$ must be rejected. The width of the deck is 4 feet.

# 9

# Algebraic Fractions

**W**orking with algebraic fractions is an elemental part of many mathematical expressions and functions. As you begin your review of this topic, take the ten-question *Benchmark Quiz* to find out what you really need to freshen up on. The questions that appear here are similar to questions that you will find on those critical exams and will give you a good indication of where your strengths lie. Check the answer key with thorough answer explanations at the completion of the chapter. Your Benchmark Quiz analysis will help you determine how much time you need to spend working with algebraic fractions and the types of problems they are contained in.

## BENCHMARK QUIZ

Unless otherwise specified, assume that no denominator equals zero.

1. Evaluate $\frac{a^3 - b}{a + b}$ for $a = -1$ and $b = 3$.
   a. $-2$
   b. $-1$
   c. $1$
   d. $2$
   e. $7$

2. For what value(s) of $x$ is the expression $\frac{x^2 - 9}{x + 2}$ undefined?
   a. $-9$
   b. $-3$
   c. $-2$
   d. $2$
   e. $3$

3. Simplify the expression $\frac{8x - 8}{4}$.
   a. $2x - 8$
   b. $2x - 4$
   c. $2x - 2$
   d. $4x - 8$
   e. $4x - 4$

4. Simplify the expression $\frac{x^2 + 3x - 10}{x + 5}$.
   a. $x^2 + 3x - 2$
   b. $x - 2$
   c. $x + 2$
   d. $x^2 + 2x - 2$
   e. $2x - 2$

5. What is the sum of the expression $\frac{2x - 5}{x - 2} + \frac{x}{2x - 4}$ in simplest form?
   a. $\frac{5}{2}$
   b. $\frac{3x - 5}{x - 2}$
   c. $\frac{x - 5}{x - 2}$
   d. $\frac{3x - 5}{2x - 4}$
   e. $\frac{4x - 5}{2x - 4}$

**6.** Subtract the expressions: $\frac{a-b}{2ab} - \frac{6}{2a}$.

   **a.** $\frac{a-b}{2a}$

   **b.** $\frac{a-6b}{2ab}$

   **c.** $\frac{a-7b}{ab}$

   **d.** $\frac{a-7b}{2ab}$

   **e.** $\frac{6a-b}{2ab}$

**7.** Find the product: $\frac{6x^3y}{5z} \cdot \frac{10z^4}{9x^3y^2}$.

   **a.** $12z^4$

   **b.** $\frac{4z^4}{3}$

   **c.** $\frac{4z^4}{3y}$

   **d.** $\frac{4z^3}{3y}$

   **e.** $\frac{2z^3}{3y^2}$

**8.** Divide: $\frac{x^2-xy}{x-y} \div \frac{6x-6y}{(x-y)^2}$.

   **a.** $\frac{x^2-xy}{6}$

   **b.** $\frac{6x}{x-y}$

   **c.** $\frac{6x}{(x-y)^2}$

   **d.** $\frac{6}{x^2-xy}$

   **e.** $6x$

**9.** Which of the following represents the complex fraction $\dfrac{\frac{1}{y}+1}{\frac{3}{y^2}}$ in simplest form?

   **a.** $\frac{3}{y}$

   **b.** $\frac{y^2+y}{3}$

   **c.** $\frac{y+1}{3}$

   **d.** $y^2 + 3y$

   **e.** $3y^2 + y$

**10.** Solve for $a$: $\frac{2}{a} = \frac{a-1}{a+3} - \frac{12}{a^2+3a}$.

    **a.** 0 or –3

    **b.** 6

    **c.** 6 or –6

    **d.** 1

    **e.** –3

## BENCHMARK QUIZ SOLUTIONS

What did you remember about working with algebraic fractions? Check your solutions against the answer explanations, and then use your Benchmark Quiz analysis to plan your study of the rest of the chapter.

### ▶ Answers

**1. a.** First substitute $a = -1$ and $b = 3$ into the expression; $\frac{(-1)^3 - 3}{-1 + 3}$. Evaluate the numerator and denominator separately. Remember that the fraction bar serves as a grouping symbol. In the numerator, evaluate the exponent first and then subtract. The numerator becomes $-1 - 3$ which is equal to $-4$. Add in the denominator to get a result of 2. The fraction is now $\frac{-4}{2}$ which simplifies to $-2$.

**2. c.** A rational expression is undefined when the denominator is equal to zero. Set the denominator equal to zero and solve; $x + 2 = 0$; $x = -2$. Therefore, the expression is undefined when $x = -2$.

**3. c.** Factor the numerator to make the expression $\frac{8(x-1)}{4}$. Divide the factor of 8 in the numerator by 4 from the denominator to simplify the expression to $\frac{2(x-1)}{1} = 2(x-1)$. Since this is not an answer choice, use the distributive property to find an equivalent form of the answer; $2x - 2$.

**4. b.** Factor the numerator to get the expression $\frac{(x+5)(x-2)}{x+5}$. Cross-cancel the common factor of $x + 5$ from the numerator and denominator and the simplified expression is $x - 2$.

**5. a.** First find the least common denominator by factoring the denominators of the expressions. $\frac{2x-5}{x-2} + \frac{x}{2(x-2)}$. The least common denominator is $2(x - 2)$. To combine the two expressions, multiply the first expression by 2 in both numerator and denominator to get a denominator of $2(x - 2)$. The second fraction already has this denominator; $\frac{2(2x-5)}{2(x-2)} + \frac{x}{2(x-2)} = \frac{4x-10}{2(x-2)} + \frac{x}{2(x-2)}$. Now add the numerators and keep the common denominator; $\frac{4x-10+x}{2(x-2)} = \frac{5x-10}{2(x-2)}$. Factor the expression completely to see if it can be simplified. The numerator factors to $5(x - 2)$, so there is a common factor of $x - 2$. The expression $\frac{5(x-2)}{2(x-2)}$ reduces to $\frac{5}{2}$.

**6. d.** The least common denominator is $2ab$, so multiply the term being subtracted by $b$ in both numerator and denominator; $\frac{a-b}{2ab} - \frac{6b}{2ab}$. Since there is now a common denominator, subtract the numerators by combining like terms; $\frac{a-b-6b}{2ab} = \frac{a-7b}{2ab}$.

**7. d.** Cross-cancel any common factors between numerators and denominators.

$$\frac{\overset{2}{\cancel{6}}\overset{3}{\cancel{x}}\overset{1}{\cancel{y}}}{\underset{1}{\cancel{5}}\underset{1}{\cancel{z}}} \cdot \frac{\overset{2}{\cancel{10}}\overset{3}{\cancel{x}}}{\underset{3}{\cancel{9}}\overset{}{\cancel{x}}\underset{1}{\overset{2}{\cancel{y}}}} = \frac{2}{1} \cdot \frac{2z^3}{3y}$$

Then multiply the remaining factors across. The result is $\frac{4z^3}{3y}$.

**8. a.** Factor each numerator and denominator completely; $\frac{x(x-y)}{x-y} \div \frac{6(x-y)}{(x-y)(x-y)}$. Change the division to multiplication and take the reciprocal of the expression being divided by; $\frac{x(x-y)}{x-y} \cdot \frac{(x-y)(x-y)}{6(x-y)}$. Cross-cancel any common factors between numerators and denominators.

$$\frac{x\,\cancel{(x-y)}}{\cancel{x-y}} \cdot \frac{(x-y)\,\cancel{(x-y)}}{6\,\cancel{(x-y)}} = \frac{x}{1} \cdot \frac{(x-y)}{6}$$

Then multiply the remaining factors across to get $\frac{x(x-y)}{6}$. Since this is not an answer choice, use the distributive property in the numerator to find an equivalent expression; $\frac{x^2-xy}{6}$.

**9. b.** The least common denominator of the terms within the complex fraction is $y^2$. Multiply each term individually by $y^2$ to reduce; $\frac{\frac{1}{y} \cdot y^2 + 1 \cdot y^2}{\frac{3}{y^2} \cdot y^2}$.

This simplifies to $\frac{y+y^2}{3}$ which is equivalent to $\frac{y^2+y}{3}$.

**10. b.** Multiply each term by the least common denominator to eliminate the fractions. The LCD is $a(a + 3)$; $\frac{2}{a} \cdot a(a + 3) = \frac{a - 1}{a + 3} \cdot a(a + 3) - \frac{12}{a(a + 3)} \cdot a(a + 3)$. After cross-canceling, the equation becomes $2(a + 3) = a(a - 1) - 12$. This simplifies to $2a + 6 = a^2 - a - 12$. Subtract $2a$ and 6 from both sides of the equation to get $a^2 - 3a - 18 = 0$. This factors to $(a - 6)(a + 3) = 0$. Set each factor equal to zero and solve for $a$ to get $a = 6$ or $a = -3$. The solution of $-3$ does not check because the denominator from the original equation cannot equal zero. Therefore the solution is 6.

## BENCHMARK QUIZ RESULTS

If you answered 8–10 questions correctly, you remember a lot about working with algebraic fractions and using those skills to solve problems. Use the following lesson to reinforce those skills and focus on any unclear concepts. The quiz at the end of the chapter will provide even more information on your strengths and full comprehension of the content.

If you answered 4–7 questions correctly, more practice and explanation in this area would be a benefit to you. Take a look at the chapter and carefully read through the skill building help. Remember that the sidebars are there to help you with hints and shortcuts. Use the quiz at the end of the chapter to assess your progress through algebraic fractions.

If you answered 1–3 questions correctly, more thorough review and practice in this area is needed. Begin with a focused read-through of the chapter, concentrating on the basics of algebraic fractions. Although much of the content is first presented at the high school level, it is easy to forget information if it is not used on a regular basis. Use the sidebars as tips to assist you in your study and then take the chapter quiz to assess your progress. Don't worry; the answer explanations are always there to help. At this point you may also want to reference more in-depth materials, such as Learning-Express's *Algebra Success in 20 Minutes a Day*.

## JUST IN TIME LESSON—ALGEBRAIC FRACTIONS

In this chapter, the skills and strategies necessary for working with algebraic fractions will be demonstrated and discussed. In particular, the topics presented will be:

- simplifying algebraic fractions
- operations with algebraic fractions
- solving equations with algebraic fractions

## ▶ Algebraic Fractions

A fraction is a ratio of two numbers, where the top number is the *numerator* and the bottom number is the *denominator*. The formal, mathematical term for the set of fractions is *rational numbers*.

## ◖◗◖◗GLOSSARY

**RATIONAL NUMBER** a number or expression that can be expressed as $\frac{a}{b}$, where $b$ is not equal to zero

Since division by zero is undefined, it is important to know when a rational expression is undefined. The fraction $\frac{5}{x-1}$ is undefined when the denominator $x - 1 = 0$, therefore $x \neq 1$. In a fraction such as $\frac{x-4}{x+3}$ the denominator would be equal to zero when $x = -3$, so this expression is undefined when $x = -3$. Notice that it really does not make a difference what is located in the numerator. Only look at the denominator to determine when a rational number is undefined.

## ▶ Using Order of Operations with Rational Expressions

Order of operations is used the same way with rational expressions as it is with simplifying other types of problems. Remember, however, to treat the division bar as a grouping symbol. For example, to simplify the expression $\frac{2x-y}{x+3}$ for $x = -4$ and $y = -1$ first substitute the values of $x$ and $y$. The expression becomes $\frac{2(-4)-(-1)}{-4+3}$. Since the fraction bar acts as a grouping symbol, evaluate the numerator and denominator separately using order of operations. The numerator becomes $-8 + 1$, which is equal to $-7$. The denominator becomes $-1$. This expression simplifies to $\frac{-7}{-1}$ which is equal to 7.

## ▶ Simplifying Algebraic Fractions

A fraction is in simplest form if the numerator and denominator of the fraction do not have any common factors, other than 1. In order to reduce fractions to simplest form where the value of the variables is unknown, find the greatest common factor of both numerator and denominator. Divide each part of the fraction by this common factor and the result is a reduced fraction. Remember that when you divide both the top and bottom of a fraction by the same value, it is the same as dividing by 1.

## ◖◗◖◗GLOSSARY

**RELATIVELY PRIME** two or more numbers or expressions whose greatest common factor is 1

When a fraction is in reduced form, the two remaining numbers or expressions in the fraction are *relatively prime*.

**1.** $\frac{6}{9} = \frac{2}{3}$               Divide numerator and denominator each by 3.

**2.** $\frac{32x}{4xy} = \frac{8}{y}$               Divide numerator and denominator each by $4x$.

Another way to look at dividing out common factors is to think of the factors as canceling out, as in the following examples.

**3.** $\frac{x^2b}{x^3b^2} = \frac{\cancel{x^2}\cancel{b}}{x^{\cancel{3}1}b^{\cancel{2}1}} = \frac{1}{xb}$ Divide numerator and denominator by the GCF of $x^2b$. Note how the factors $x^2$ and $b$ cancel out in the fraction.

**4.** $\frac{x^2-9}{3x-9} = \frac{(x+3)(x\cancel{-3})}{3(x\cancel{-3})} = \frac{(x+3)}{3}$

First factor the expressions in the numerator and denominator. Then divide (cancel) the common factors. In this case the common factor was $x - 3$.

## ▶ More Reducing with Polynomials

When simplifying a rational expression with a polynomial in the denominator, another approach is to perform long division. This process looks very much like long division with numbers. The following example uses long division to reduce a fraction.

$\dfrac{x^2+8x-33}{x-3}$ means $(x^2+8x-33) \div (x-3)$

**Set up long division:**

$$\begin{array}{r} x \phantom{+8x-33} \\ x-3\overline{\smash{\big)}\,x^2+8x-33} \\ x^2-3x \phantom{-33} \end{array}$$

Divide $x^2$ by $x$ to get $x$ on top. Multiply this $x$ by $x - 3$ and place under $x^2 + 8x$.

➡

$$\begin{array}{r} x \phantom{+8x-33} \\ x-3\overline{\smash{\big)}\,x^2+8x-33} \\ -x^2-3x \phantom{-33} \\ \hline 11x-33 \end{array}$$

Subtract like to get $11x$. Bring down the $-33$ and repeat the division process.

➡

$$\begin{array}{r} x+11 \phantom{-3} \\ x-3\overline{\smash{\big)}\,x^2+8x-33} \\ -x^2+x \phantom{-33} \\ \hline 11x-33 \\ 11x-33 \end{array}$$

$11x$ divided by $x$ is equal to 11. Place the $+11$ on top after the $x$ and multiply $x - 3$ by 11. Place this product under $11x - 33$.

➡

$$\begin{array}{r} x+11 \phantom{-3} \\ x-3\overline{\smash{\big)}\,x^2+8x-33} \\ -x^2+3x \phantom{-33} \\ \hline 11x-33 \\ -11x+33 \\ \hline 0 \end{array}$$

Subtract like terms. Since the remainder is zero, the polynomial was factorable. $x + 11$ is the other factor.

Thus, the result is $x + 11$.

Although factoring an expression and canceling the common factors may be quicker, this method is a good choice if the factors are difficult to find or if the polynomial in the numerator is not factorable.

When performing operations with fractions, it is important to remember the cases where you need a common denominator and when a common denominator is not necessary.

## ▶ Adding and Subtracting Algebraic Fractions

It is very important to find the least common denominator, LCD, when adding or subtracting fractions. To find a common denominator, find the least common multiple of the denominators in the expressions. Then rewrite each fraction using this common multiple as the denominator. After this is done, you will only be adding or subtracting the numerators and keeping the common denominator as the bottom number in your answer. Here are a few examples.

1. $\frac{1}{x} + \frac{7}{x} = \frac{8}{x}$

   These already have a common denominator of $x$, so just add the numerators and keep the denominator.

2. $\frac{2}{5} + \frac{2}{3}$

   LCD = 15

   $\frac{2 \cdot 3}{5 \cdot 3} + \frac{2 \cdot 5}{3 \cdot 5}$

   $\frac{6}{15} + \frac{10}{15} = \frac{16}{15}$

Multiply the numerator and denominator of the first fraction by 3 and the numerator and denominator of the second fraction by 5 to get a common denominator of 15 before adding the numerators. Remember that when you multiply both the numerator and denominator by the same value, you are really just multiplying the fraction by 1. For example, $\frac{3}{3}$ is equal to 1.

 RULE BOOK

**Be sure to only cancel factors when reducing rational expressions. Factors are connected to everything else by multiplication. You may not cancel terms that are connected by addition or subtraction.**

3. $\frac{3}{y} + \frac{4}{xy}$

   LCD = $xy$

   $\frac{3 \cdot x}{y \cdot x} + \frac{4}{xy} = \frac{3x + 4}{xy}$

Multiply the numerator and denominator of the first fraction by $x$ to get a common denominator of $xy$ before adding the numerators. After adding the numerators, the expression is in simplest form.

**4.** $\frac{1}{x-1} + \frac{3}{x}$

LCD $= x(x-1)$

$\frac{1 \cdot x}{(x-1) \cdot x} + \frac{3 \cdot (x-1)}{x \cdot (x-1)} = \frac{x + 3x - 3}{x(x-1)} = \frac{4x-3}{x(x-1)}$

Multiply the numerator and denominator of the first fraction by $x$ and the numerator and denominator of the second fraction by $x-1$ to get a common denominator of $x(x-1)$ before combining them.

**5.** $\frac{x+6}{x} - \frac{x-2}{3x}$

LCD $= 3x$

$\frac{(x+6) \cdot 3}{x \cdot 3} - \frac{(x-2)}{3x} = \frac{3x + 18 - x + 2}{3x} = \frac{2x + 20}{3x}$

Multiply the numerator and denominator of the first fraction by 3 and distribute the subtraction to both terms in the numerator of the second fraction before combining them.

 RULE BOOK

**Be careful to distribute the subtraction sign to ALL terms in an expression that is being subtracted.**

## ▶ Multiplying Algebraic Fractions

It is not necessary to get a common denominator when multiplying fractions. To perform this operation you can simply multiply across the numerators and then denominators. If possible, you may first cross-cancel common factors if they are present, as in examples **b** and **c**.

**1.** $\frac{1}{3} \cdot \frac{2}{3} = \frac{2}{9}$

There are no common factors so just multiply across.

**2.**  $\frac{12}{25} \cdot \frac{5}{3} = \frac{\overset{4}{\cancel{12}}}{\underset{5}{\cancel{25}}} \cdot \frac{\overset{1}{\cancel{5}}}{\underset{1}{\cancel{3}}} = \frac{4}{5}$

Cancel out the common factors of 4 and 5 and then multiply across.

 RULE BOOK

**When performing operations with rational expressions always Factor First, if possible.**

**3.** $\dfrac{4x}{x^2-16} \bullet \dfrac{x+4}{2x^2} = \dfrac{\overset{2}{\cancel{4}}\cancel{x}(x\cancel{+4})}{\cancel{2}x\overset{1}{\cancel{^2}}(x-4)(x\cancel{+4})} = \dfrac{2}{x(x-4)}$

In this example, factor each part of the expressions first and then cancel out the common factors of 2, $x$, and $x + 4$.

## RULE BOOK

When multiplying and dividing rational numbers, it is necessary to cross-cancel a factor from the *numerator* with a factor from the *denominator* when finding pairs of factors to cancel out.

# ▶ Dividing Algebraic Fractions

A common denominator is also not needed when dividing fractions and the procedure is similar to multiplying. Since dividing by a fraction is the same as multiplying by its reciprocal, leave the first fraction as it is, change the division symbol to multiplication and change the number being divided by to its reciprocal. Here are a few examples of dividing rational expressions.

**1.** $\dfrac{4}{5} \div \dfrac{4}{3} = \dfrac{\overset{1}{\cancel{4}}}{5} \bullet \dfrac{3}{\underset{1}{\cancel{4}}} = \dfrac{3}{5}$

Take the reciprocal of $\frac{4}{3}$ and cancel the common factor of 4 when multiplying.

**2.** $\dfrac{3x}{y} \div \dfrac{12x}{5xy} = \dfrac{\overset{1\,1}{\cancel{3}\cancel{x}}}{\underset{1}{\cancel{y}}} \bullet \dfrac{\overset{1}{5x\cancel{y}}}{\underset{4\,1}{\cancel{12}\cancel{x}}} = \dfrac{5x}{4}$

Take the reciprocal of the second fraction, cross-cancel the common factors, and multiply the remaining values.

**3.** $\dfrac{a^2+2a}{a^2+3a+2} \div \dfrac{a^2-3a}{2a+2} = \dfrac{\cancel{a}(a\cancel{+2})}{(a\cancel{+1})(a\cancel{+2})} \bullet \dfrac{2(a\cancel{+1})}{\cancel{a}(a-3)} = \dfrac{2}{a-3}$

Remember to **F**actor **F**irst before cross-canceling the factors of $a$, $a + 1$, and $a + 2$.

## RULE BOOK

When adding or subtracting rational expressions (fractions), a common denominator is needed. When multiplying and dividing rationals a common denominator is **not** needed.

# ▶ Simplifying Complex Fractions

Complex fractions are rational expressions where the numerator and/or denominator is also a fraction. These are fractions within fractions. Although they look complicated, as the name seems to imply, they are not difficult to simplify. In order to simplify a complex fraction, find the least common denominator of all of the smaller fractions within the complex fraction. Then multiply each term individually by the common denominator. The fraction should not be complex after this step. Take the following example.

Simplify the complex fraction $\frac{\frac{x}{2} + \frac{1}{x^2}}{\frac{x+1}{x}}$.

Within the complex fraction, there are 3 terms: $\frac{x}{2}$, $\frac{1}{x^2}$, and $\frac{x+1}{x}$. The least common denominator of these three fractions is $2x^2$, so multiply each term by $2x^2$. The fraction would look like this:

$$\frac{\frac{x}{2} \cdot 2x^2 + \frac{1}{x^2} \cdot 2x^2}{\frac{x+1}{x} \cdot 2x^2}.$$

After cross canceling, the complex fraction would then become $\frac{x^3 + 2}{2x^2 + 2x}$ which is in simplest form.

## ▶ Solving Equations with Algebraic Fractions

When solving equations that contain fractions, very often the easiest method is to first figure out the least common multiple of the denominators in the equation. Then multiply each individual term of the equation by this least common denominator, or LCD. After this step, all fractions should be eliminated and the resulting equation should be much easier to solve.

**1.** Solve for $x$:    $\frac{2}{3}x + \frac{1}{6}x = \frac{1}{4}$.
   Multiply each term by the LCD = 12:
$$\left(\frac{2}{3}x \cdot 12\right) + \left(\frac{1}{6}x \cdot 12\right) = \left(\frac{1}{4} \cdot 12\right).$$
   The equation now becomes $8x + 2x = 3$.
   Combine like terms to get $10x = 3$.
   Divide each side of the equal sign by 10: $x = \frac{3}{10}$.

**2.** Solve for $x$:    $\frac{1}{x} = \frac{1}{4} + \frac{1}{6}$.
   Multiply each term by the LCD = $12x$:
$$\left(\frac{1}{x} \cdot 12x\right) = \left(\frac{1}{4} \cdot 12x\right) + \left(\frac{1}{6} \cdot 12x\right).$$
   The equation now becomes $12 = 3x + 2x$.
   Combine like terms:    $12 = 5x$.
   Divide each side by 5: $x = \frac{12}{5} = 2.4$.

**3.** Solve for $y$:    $\frac{4}{y} + \frac{3}{y-3} = \frac{y}{y-3}$
   Multiply each term by the LCD = $y(y-3)$:
$$\frac{4}{y} \cdot y(y-3) + \frac{3}{y-3} \cdot y(y-3) = \frac{y}{y-3} \cdot y(y-3).$$

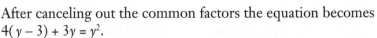

After canceling out the common factors the equation becomes $4(y - 3) + 3y = y^2$.

Use the distributive property on the left side: $4y - 12 + 3y = y^2$.

Combine like terms to get $7y - 12 = y^2$.

Put the equation in standard form: $y^2 - 7y + 12 = 0$.

Factor the equation $(y - 3)(y - 4) = 0$.

Set each factor equal to zero and solve: $y - 3 = 0$ or $y - 4 = 0$.

The result is $y = 3$ or $4$.

However, by checking the equation we realize $y$ cannot be equal to 3. If you substitute 3 in for $y$ into the original equation, the denominators of $y - 3$ will equal zero. Remember that division by zero is undefined. Therefore, the solution is 4.

 EXTRA HELP

**For more examples and explanations of working with rational expressions, check out the website www.mathnotes.com. Here you will find a lesson devoted to rational expressions, along with an array of other mathematical topics.**

## TIPS AND STRATEGIES

Fractions can be one of the more intimidating mathematical concepts and you just can't seem to avoid them. Use the following hints and shortcuts to help pave the way to a smoother road with rational expressions and equations.

- Keep in mind that rational numbers are really just fractions.
- Remember that division by zero is undefined. This means never consider a solution that makes the denominator of a fraction equal to zero.
- Simplify rationals by dividing both the numerator and denominator of the fraction by the greatest common factor, or GCF.
- Always find a common denominator when adding or subtracting rational numbers.
- When multiplying or dividing rational numbers, a common denominator is not necessary.
- Always completely factor any expression before performing any operation using the expression. Often many of the factors will cancel out before you actually do the operation.
- When cross-canceling factors in a multiplication or division problem, it is important to cancel a numerator with a denominator.
- Do not make the mistake of cross-canceling in an addition or subtraction problem; you can only cancel *factors*.

- When solving equations containing fractions, multiply the entire equation by the LCD to eliminate the fractions. Then solve the equation.
- Use the decimal form of fractions when possible, especially if the questions are multiple-choice and/or calculators are allowed.

## CHAPTER QUIZ

The following chapter quiz provides even more practice using algebraic fractions. Take it now and see how far you have come. Don't forget to use the answer explanations at the conclusion of the quiz to clarify and reinforce the important topics covered in this chapter. Unless otherwise specified, assume that no denominator equals zero.

**1.** Evaluate $\frac{a^2 + 4b}{a - b}$ when $a = -2$ and $b = 1$.

   **a.** 0

   **b.** $\frac{-8}{3}$

   **c.** $-8$

   **d.** 8

   **e.** $-4$

**2.** Evaluate $\frac{cd^2 - 2}{cd}$ for $c = 4$ and $d = -2$.

   **a.** $\frac{-17}{4}$

   **b.** $\frac{-7}{4}$

   **c.** $-4$

   **d.** $-2$

   **e.** $\frac{17}{4}$

**3.** For what value of $x$ is the expression $\frac{x^2 + x - 6}{5x}$ undefined?

   **a.** $-6$

   **b.** $-3$

   **c.** 0

   **d.** 2

   **e.** 5

**4.** For what value(s) of $x$ is the expression $\frac{3x-5}{x^2-4}$ undefined?

 **a.** $-4$
 **b.** 4 or $-4$
 **c.** 3 or $-5$
 **d.** 2
 **e.** 2 or $-2$

**5.** Simplify the expression $\frac{24c^3d - 2cd}{4cd}$.

 **a.** $6c^2$
 **b.** $\frac{6c^2}{2}$
 **c.** $12c^2$
 **d.** $\frac{12c^2 - 1}{2}$
 **e.** $6c^2 - 1$

**6.** Simplify the expression $\frac{n^2 - 25}{n + 5}$.

 **a.** $n$
 **b.** $-5$
 **c.** $5n$
 **d.** $n + 5$
 **e.** $n - 5$

**7.** Simplify the expression: $\frac{4x^2 - 12x + 8}{x^2 - 3x + 2}$.

 **a.** $x - 1$
 **b.** $x + 2$
 **c.** 4
 **d.** $4x - 1$
 **e.** $8 - 4x$

**8.** Find the sum of the expressions: $\frac{7y}{x} + \frac{10y}{x}$.

 **a.** $17y$
 **b.** $\frac{17y}{x}$
 **c.** $\frac{17y}{x^2}$
 **d.** $\frac{17y^2}{x}$
 **e.** $\frac{17y^2}{x^2}$

**9.** Find the sum of the expressions: $\frac{3}{xy} + \frac{6x}{y}$.

 **a.** $\frac{6x + 3}{x}$
 **b.** $\frac{6x + 3}{xy}$
 **c.** $\frac{6x^2 + 3}{y}$
 **d.** $\frac{6x^2 + 3}{xy}$
 **e.** $\frac{18x^2}{xy^2}$

**10.** Find the sum of the expressions: $\frac{x}{x^2-1} + \frac{2x}{x-1}$.

a. $\frac{3x}{x^2-1}$

b. $\frac{2x^2+3x}{x^2-1}$

c. $\frac{2x^2+x+1}{x^2-1}$

d. $\frac{2x^2+3x}{x-1}$

e. $\frac{2x^2+x}{x^2-1}$

**11.** Find the difference of the expressions: $\frac{2}{a} - \frac{4b}{a}$.

a. $-2b$

b. $\frac{-2b}{a}$

c. $\frac{2-4b}{a}$

d. $\frac{2-4b}{a^2}$

e. $\frac{-4b}{a}$

**12.** Find the difference of the expressions: $\frac{c}{3d} - \frac{c-6}{cd}$.

a. $\frac{-6}{3d}$

b. $\frac{-6}{3cd}$

c. $\frac{c^2-c+6}{3cd}$

d. $\frac{c^2-3c+18}{3cd}$

e. $\frac{c^2-3c+18}{3d}$

**13.** Find the product of the expressions: $\frac{12nm^3}{5x} \cdot \frac{30x^2}{48nm}$.

a. $3m^3x$

b. $\frac{3m^2x}{2}$

c. $\frac{3m^3x}{2}$

d. $\frac{6m^2x}{2}$

e. $\frac{6m^2x}{5}$

**14.** Find the product of the expressions: $\frac{x^2 + 3x}{x^2 - 49} \cdot \frac{x + 7}{x}$.

a. $\frac{x + 3}{x - 7}$

b. $\frac{x^2 + 3}{x - 7}$

c. $\frac{3x^2 + 7}{x^2 - 49}$

d. $3x^2 - 7$

e. $\frac{3x^2 + 7}{x - 7}$

**15.** Find the product of the expressions: $\frac{(a - b)^2}{x - 3} \cdot \frac{x^2 - 3x}{a - b}$.

a. $x$

b. $a - b$

c. $\frac{a - b}{x - 3}$

d. $\frac{ax - bx}{x - 3}$

e. $ax - bx$

**16.** Find the quotient of the expressions: $\frac{x^2 + 6x + 5}{3x^3} \div \frac{x + 1}{6xy}$.

a. $\frac{2xy + 10y}{x^2}$

b. $\frac{2xy + 10y}{x^3y}$

c. $\frac{x^2 + 7x + 6}{18x^4y}$

d. $\frac{2x^2y + 10y}{3x^2 + 3x}$

e. $\frac{x + 10y}{3x^2}$

**17.** Find the quotient of the expressions: $\frac{x}{y} \div \frac{4}{y} \div \frac{x^2}{12}$.

a. $\frac{x}{3}$

b. $\frac{3}{x}$

c. $\frac{x^3}{48}$

d. $\frac{48}{xy^2}$

e. $\frac{x^3}{3y^2}$

**18.** Simplify the complex fraction: $\dfrac{\frac{1}{x}}{\frac{4}{x^2}}$.

   **a.** $x$

   **b.** $4$

   **c.** $\dfrac{x}{4}$

   **d.** $\dfrac{4}{x}$

   **e.** $\dfrac{x^2}{4}$

**19.** Simplify the complex fraction: $\dfrac{\frac{1}{y}+y}{\frac{8}{y^2}}$.

   **a.** $y^3 + y$

   **b.** $\dfrac{y^2 + y}{8}$

   **c.** $\dfrac{8}{y^2 + y}$

   **d.** $y + \dfrac{y^2}{8}$

   **e.** $\dfrac{y^3 + y}{8}$

**20.** Simplify the complex fraction: $\dfrac{\frac{1}{y^2}+1}{1-\frac{1}{2y}}$.

   **a.** $\dfrac{2 + 2y^2}{2y^2}$

   **b.** $\dfrac{2 + 2y^2}{2y^2 - 1}$

   **c.** $\dfrac{1 + 2y^2}{2y^2 - y}$

   **d.** $\dfrac{2 + 2y^2}{2y^2 - y}$

   **e.** $2 + 2y^2$

**21.** Solve for $x$: $\dfrac{x}{3} = \dfrac{x}{18} + \dfrac{10}{9}$.

   **a.** $1$

   **b.** $2$

   **c.** $4$

   **d.** $5$

   **e.** $18$

**22.** Solve for $x$: $\dfrac{4}{x} - 5 = \dfrac{9}{x}$.

   **a.** $-5.6$

   **b.** $-5$

   **c.** $-1$

   **d.** $1$

   **e.** $5.6$

23. Solve for $x$: $\frac{1}{3} = \frac{a+4}{a+2}$.
  a. $-5$
  b. $-2$
  c. 3
  d. 5
  e. 7

24. Solve for $x$: $\frac{3}{x-2} - \frac{6}{x^2-4} = \frac{x}{x+2}$.
  a. $-2$
  b. $-5$
  c. 5
  d. 2 or $-2$
  e. 0 or 5

25. Solve for $x$: $\frac{-1}{x-3} + \frac{x}{3x-9} = \frac{x}{3}$.
  a. 1
  b. 1 or 3
  c. 3
  d. $-3$ or 3
  e. 0 or 3

## ANSWERS

Use the explanations written here to help identify errors and reinforce skills that you have mastered. Don't forget about the materials referenced in the text that provide even more examples and practice using algebraic fractions.

**1. b.** Substitute the values for the variables in the expression; $\frac{(-2)^2 + 4(1)}{(-2)-(1)}$. Simplify the numerator and denominator separately. In the numerator, evaluate the exponent first. Remember that $(-2)2 = (-2)(-2) = 4$. The numerator $(-2)2 + 4(1)$ becomes $(4) + 4(1)$. Since $4(1) = 4$, add $4 + 4 = 8$. In the denominator, change subtraction to addition and change the sign of the second term to its opposite; $(-2) - (1)$ then becomes $(-2) + (-1)$. The signs are the same so add and keep the sign; $-2 + -1 = -3$. Putting the fraction back together with the numerator on top and the denominator on the bottom results in the solution of $\frac{8}{-3} = \frac{-8}{3}$.

**2. b.** Substitute the values for the variables in the expression; $\frac{4(-2)^2 - 2}{(4)(-2)}$. Simplify the numerator and denominator separately. In the numerator, evaluate the exponent first. Remember that $(-2)2 =$ $(-2)(-2) = 4$. The numerator $4(-2)^2 - 2$ becomes $4(4) - 2$. Since $4(4) = 16$, subtract $16 - 2 = 14$. In the denominator, multiply 4 and $-2$ to get a result of $-8$. Putting the fraction back together with the numerator on top and the denominator on the bottom results in the solution of $\frac{14}{-8}$, which reduces to $\frac{-7}{4}$.

**3. c.** The expression is undefined when the denominator is equal to zero. Set the denominator equal to zero and solve for $x$; $5x = 0$. Divide both sides of the equal sign by 5 to get $x = 0$. Therefore, when $x = 0$ the expression is undefined.

**4. e.** The expression is undefined when the denominator is equal to zero. Set the denominator equal to zero and solve for $x$; $x^2 - 4 = 0$. Since this binomial is the difference between 2 squares, factor the left side to $(x - 2)(x + 2)$ and set each factor equal to zero. The result is $x = 2$ or $x = -2$. Therefore, when $x = 2$ or $-2$ the expression is undefined.

**5. d.** To simplify, factor the numerator of the expression and look for any common factors between the numerator and denominator. When factored, the expression becomes $\frac{2cd(12c^2 - 1)}{4cd}$. Cancel the common factor of $2cd$ from both the numerator and denominator to get the expression $\frac{12c^2 - 1}{2}$.

**6. e.** To simplify, factor the numerator of the expression and look for common factors between the numerator and denominator. The binomial $n^2 - 25$ factors to $(n - 5)(n + 5)$. In the expression there is a common factor of $n + 5$. After canceling these factors the expression simplifies to $n - 5$.

**7. c.** Factor both the numerator and denominator and look for common factors to cancel out. Factoring each trinomial, the expression becomes $\frac{4(x - 1)(x - 2)}{(x - 1)(x - 2)}$. Canceling the common factors of $x - 1$ and $x - 2$ leaves a simplified answer of 4.

**8. b.** Since there is already a common denominator of $x$, add the numerators and keep the denominator; $\frac{7y}{x} + \frac{10y}{x}$ becomes $\frac{7y + 10y}{x}$ which equals $\frac{17y}{x}$.

**9. d.** Find the least common denominator between the fractions and then add the numerators. The LCD is $xy$, so multiply the second fraction by $x$ in the numerator and denominator; $\frac{3}{xy} + \frac{6x}{y} = \frac{3}{xy} + \frac{6x \cdot x}{y \cdot x}$ $= \frac{3}{xy} + \frac{6x^2}{xy}$. Add the numerators and keep the denominator; $\frac{3 + 6x^2}{xy}$ which is equivalent to answer choice **d.**

**10. b.** Since the denominator of the first fraction factors to $(x - 1)(x + 1)$, examine the denominators to find that the least common denominator is $(x - 1)(x + 1)$. Multiply the second fraction by $x + 1$ to convert to this denominator; $\frac{x}{x^2 - 1} + \frac{2 \cdot (x + 1)}{(x - 1) \cdot (x + 1)} = \frac{x}{(x - 1)(x + 1)}$ $+ \frac{2x^2 + 2x}{(x - 1)(x + 1)}$. Add the numerators; $\frac{2x^2 + 3x}{(x - 1)(x + 1)} = \frac{2x^2 + 3x}{x^2 - 1}$.

**11. c.** Since the fractions already have a common denominator, subtract the numerators and keep the denominator; $\frac{2 - 4b}{a}$.

**12. d.** The least common denominator of the fractions is $3cd$. Multiply the first fraction by $c$ and the second fraction by 3 in the numerator and denominator to convert each to the common denominator; $\frac{c \cdot c}{3d \cdot c} - \frac{(c - 6) \cdot 3}{cd \cdot 3} = \frac{c^2}{3cd} - \frac{3c - 18}{3cd}$. Subtract the numerators. Keep in mind that if there is more than one term being subtracted, as there is in this case, be sure to subtract each term. The expression becomes $\frac{c^2 - 3c + 18}{3cd}$.

**13. b.** Cross-cancel the common factors between the numerators and denominators; $\frac{12mm^3}{5x} \cdot \frac{30x^2}{48nm}$ becomes $\frac{m^2}{1} \cdot \frac{3x}{2}$. Multiply across to get the answer of $\frac{3m^2x}{2}$.

**14. a.** Completely factor each numerator and denominator; $\frac{x^2 + 3x}{x^2 - 49} \cdot \frac{x + 7}{x}$ becomes $\frac{x(x + 3)}{(x - 7)(x + 7)} \cdot \frac{x + 7}{x}$. Cross-cancel the common factors of $x$ and $x + 7$. Multiply the remaining factors across to get $\frac{x + 3}{x - 7}$.

**15. e.** Factor each part of the expressions completely; $\frac{(a - b)^2}{x - 3} \cdot \frac{x^2 - 3x}{a - b} = \frac{(a - b)(a - b)}{x - 3} \cdot \frac{x(x - 3)}{a - b}$. Cross-cancel the common factors of $a - b$ and $x - 3$. The result is $(a - b) \cdot x$. The result is $ax - bx$.

**16. a.** Factor each part of the expressions completely; $\dfrac{x^2 + 6x + 5}{3x^3} \div \dfrac{x + 1}{6xy} = \dfrac{(x + 1)(x + 5)}{3x^3} \div \dfrac{x + 1}{6xy}$. Change division to multiplication and the fraction being divided to its reciprocal; $\dfrac{(x + 1)(x + 5)}{3x^3} \bullet \dfrac{6xy}{x + 1}$. Cross-cancel the common factors of $x + 1$, 3, and $x$. The expressions become $\dfrac{x + 5}{x^2} \bullet \dfrac{2y}{1}$ which multiplies to $\dfrac{2xy + 10y}{x^2}$.

**17. b.** Divide the first two expressions by multiplying by the reciprocal of the second fraction; $\dfrac{x}{y} \bullet \dfrac{y}{4}$. Since the factors of $y$ cancel out the result is $\dfrac{x}{4}$. Now divide this result by the third fraction by multiplying by its reciprocal; $\dfrac{x}{4} \bullet \dfrac{12}{x^2}$. Cross-cancel the factors of $x$ and 4 to get a result of $\dfrac{3}{x}$.

**18. c.** Since the least common denominator of both fractions within the complex fraction is $x^2$, multiply each of these by $x^2$ and cross-cancel;

$$\dfrac{\frac{1}{x} \bullet x^2}{\frac{4}{x^2} \bullet x^2} = \dfrac{x}{4}.$$

**19. e.** Since the least common denominator of both fractions within the complex fraction is $y^2$, multiply each of these by $y^2$ and cross-cancel;

$$\dfrac{\frac{1}{y} \bullet y^2 + y \bullet y^2}{\frac{8}{y^2} \bullet y^2} = \dfrac{y + y^3}{8}.$$

**20. d.** Since the least common denominator of both fractions within the complex fraction is $2y^2$, multiply each of these by $2y^2$ and cross-cancel;

$$\dfrac{\frac{1}{y^2} \bullet 2y^2 + 1 \bullet 2y^2}{1 \bullet 2y^2 - \frac{1}{2y} \bullet 2y^2} = \dfrac{2 + 2y^2}{2y^2 - y}.$$

**21. c.** Multiply each fraction in the equation by the least common denominator of 18; $\dfrac{x}{3} \bullet 18 = \dfrac{x}{18} \bullet 18 + \dfrac{10}{9} \bullet 18$. The equation now reduces to $6x = x + 20$. Subtract $x$ from both sides of the equal sign to get $5x = 20$. Divide both sides by 5; $x = 4$.

**22. c.** Multiply each fraction in the equation by the least common denominator of $x$; $\dfrac{4}{x} \bullet x - 5 \bullet x = \dfrac{9}{x} \bullet x$. The equation simplifies to $4 - 5x = 9$. Subtract 4 from both sides of the equal sign; $-5x = 5$. Divide both sides of the equation by $-5$; $x = -1$.

**23. a.** This question is a special case equation that can be solved by cross-multiplying. This gives an equation of $a + 2 = 3a + 12$. (Don't forget to use distributive property.) Subtract $a$ from both sides of the equal sign to get $2 = 2a + 12$. Subtract 12 from both sides; $-10 = 2a$. Divide both sides by 2 to get $-5 = a$.

**24. e.** Multiply each fraction in the equation by the least common denominator of $(x - 2)(x + 2)$ which is equivalent to $x^2 - 4$; $\frac{3}{x-2}$ · $(x - 2)(x + 2) - \frac{6}{(x-2)(x+2)} \cdot (x - 2)(x + 2) = \frac{x}{x+2} \cdot (x - 2)(x + 2)$. The equation simplifies to $3(x + 2) - 6 = x(x - 2)$. Use distributive property to eliminate the parentheses; $3x + 6 - 6 = x^2 - 2x$. Simplify; $3x = x^2 - 2x$. Subtract $3x$ from both sides of the equal sign; $0 = x^2 - 5x$. Factor the right side of the equation and set each factor equal to zero; $0 = x(x - 5)$; $x = 0$ or $x - 5 = 0$. Therefore, the solution is $x = 0$ or 5.

**25. a.** Multiply each fraction in the equation by the least common denominator of $3(x - 3)$, which is equivalent to $3x - 9$; $\frac{-1}{x-3} \cdot 3(x - 3) + \frac{x}{3(x-3)} \cdot 3(x - 3) = \frac{x}{3} \cdot 3(x - 3)$. The equation simplifies to $-3 + x = x(x - 3)$. Use the distributive property to eliminate the parentheses; $-3 + x = x^2 - 3x$. Add 3 and subtract $x$ from both sides of the equal sign to get the equation in standard form; $x^2 - 4x + 3 = 0$. Factor the left side of the equation; $(x - 1)(x - 3) = 0$. Set each factor equal to zero and solve for $x$; $x - 1 = 0$ or $x - 3 = 0$; $x = 1$ or 3. When 3 is substituted into the original equation for $x$, the denominator of $x - 3$ is equal to zero, or undefined. Therefore, the solution is 1.

# Translating Algebraic Expressions and Solving Word Problems

**A**s you start your review of the basics of translating algebraic expressions and solving word problems, commit a few minutes to this ten-question *Benchmark Quiz*. The questions you will find here are similar to the type of questions that you may encounter on important tests. When it is completed, use the answer explanations that follow the quiz to check your results. This Benchmark Quiz analysis will help you determine what type(s) of word problems you are confident with already and the type you may want to spend your review and practice time on.

## BENCHMARK QUIZ

1. Which of the following expressions is equivalent to "Ten more than three times a number"?
   **a.** $\frac{x}{3} + 10$
   **b.** $3x - 10$
   **c.** $3(x + 10)$
   **d.** $3x + 10$
   **e.** $3x + 10x$

2. Three times the sum of a number and 5 is equal to 27. What is the number?
   a. 3
   b. 4
   c. 7
   d. 21
   e. 22

3. One integer is four times another. The sum of the integers is 5. What is the value of the lesser integer?
   a. 5
   b. 4
   c. 2
   d. 1
   e. 0

4. The sum of three times the greater integer and 5 times the lesser integer is 9. Three less than the greater equals the lesser. What is the value of the lesser integer?
   a. 0
   b. 1
   c. 2
   d. 3
   e. 9

5. The perimeter of a rectangle is 104 inches. The width is 6 inches less than 3 times the length. Find the width of the rectangle.
   a. 13.5
   b. 14.5
   c. 15
   d. 17
   e. 37.5

6. What is the greater of two consecutive positive odd integers whose sum is 256?
   a. 16
   b. 32
   c. 33
   d. 129
   e. 131

7. What is the lesser of two consecutive positive even integers whose product is 1,088?
   a. 22
   b. 32
   c. 34
   d. 36
   e. 33

8. Kala invested $2,000 in an account that earns 5% interest and $x$ amount in a different account that earns 7% interest. How much is invested at 7% if the total amount of interest earned is $310?
   a. $2,000.00
   b. $2,070.00
   c. $2,068.75
   d. $3,000.00
   e. $5,000.00

9. Sheila bought 10 CDs that cost $d$ dollars each. What is the total cost of the CDs in terms of $d$?
   a. $d + 10$
   b. $d - 10$
   c. $10d$
   d. $\frac{d}{10}$
   e. $\frac{10}{d}$

10. Nancy and Jack can shovel the driveway together in 6 hours. If it takes Nancy 10 hours working alone, how long will it take Jack working alone?
    a. 4
    b. 6
    c. 10
    d. 15
    e. 16

## BENCHMARK QUIZ SOLUTIONS

How did you do translating and solving word problems? Compare your answers with the explanations here and then make your plan for careful review and practice.

## ▶ *Answers*

**1. d.** The key words *more than* tell you to use addition and *three times a number* translates to $3x$. Therefore, ten more than three times a number translates to the expression $3x + 10$.

**2. b.** Let $x$ = a number. The part of the sentence "three times the sum of a number and 5" translates to $3(x + 5)$. The second part of the sentence sets this equal to 27. The equation is $3(x + 5) = 27$. Use the distributive property on the left side of the equal sign; $3x + 15 = 27$. Subtract 15 from both sides of the equation; $3x + 15 - 15 = 27 - 15$. This simplifies to $3x = 12$. Divide both sides by 3; $\frac{3x}{3} = \frac{12}{3}$. Therefore the number is 4.

**3. d.** Let $x$ = the lesser integer and let $y$ = the greater integer. The first sentence in the question gives the equation $y = 4x$. The second sentence gives the equation $x + y = 5$. Substitute $y = 4x$ into the second equation; $x + 4x = 5$. Combine like terms on the left side of the equation; $5x = 5$. Divide both sides of the equation by 5; $\frac{5x}{5} = \frac{5}{5}$. This gives a solution of $x = 1$, which is the lesser integer.

**4. a.** Let $x$ = the lesser integer and let $y$ = the greater integer. The first sentence in the question gives the equation $3y + 5x = 9$. The second sentence gives the equation $y - 3 = x$. Substitute $y - 3$ for $x$ in the first equation; $3y + 5(y - 3) = 9$. Use the distributive property on the left side of the equation; $3y + 5y - 15 = 9$. Combine like terms on the left side; $8y - 15 = 9$. Add 15 to both sides of the equation; $8y - 15 + 15 = 9 + 15$. Divide both sides of the equation by 8; $\frac{8y}{8} = \frac{24}{8}$. This gives a solution of $y = 3$. Therefore the lesser, $x$, is three less than $y$, so $x = 0$.

**5. e.** Let $l$ = the length of the rectangle and let $w$ = the width of the rectangle. Since the width is 6 inches less than 3 times the length, one equation is $w = 3l - 6$. The formula for the perimeter of a rectangle is $2l + 2w = 104$. Substituting the first equation into the perimeter equation for $w$ results in $2l + 2(3l - 6) = 104$. Use the distributive property on the left side of the equation; $2l + 6l - 12 = 104$. Combine like terms on the left side of the equation; $8l - 12 = 104$. Add 12 to both sides of the equation; $8l - 12 + 12 = 104 + 12$. Divide both sides of the equation by 8; $\frac{8l}{8} = \frac{116}{8}$. Therefore, the length is $l = 14.5$ inches and the width is $w = 3(14.5) - 6 = 37.5$ inches.

**6. d.** Let $x$ = the lesser odd integer and let $x + 2$ = the greater odd integer. Since *sum* is a key word for addition, the equation is $x + x + 2 = 256$. Combine like terms on the left side of the equation; $2x + 2 = 256$. Subtract 2 from both sides of the equation; $2x + 2 - 2 = 256 - 2$. This simplifies to $2x = 254$. Divide each side by 2; $\frac{2x}{2} = \frac{254}{2}$; $x = 127$. Thus, the greater odd integer is $x + 2 = 129$.

**7. b.** Let $x$ = the lesser even integer and let $x + 2$ = the greater even integer. Since *product* is a key word for multiplication, the equation is $x(x + 2) = 1{,}088$. Multiply using the distributive property on the left side of the equation; $x^2 + 2x = 1{,}088$. Put the equation in standard form and set it equal to zero; $x^2 + 2x - 1{,}088 = 0$. Factor the trinomial; $(x - 32)(x + 34) = 0$. **Tip:** Since it is difficult to find the factors of 1,088, use the possible answers to help you guess and check. Set each factor equal to zero and solve; $x - 32 = 0$ or $x + 34 = 0$. So, $x = 32$ or $x = -34$. Since you are looking for a positive integer, reject the $x$-value of $-34$. Therefore, the lesser positive integer would be 32.

**8. d.** Let $x$ = the amount invested at 7% interest. Since the total interest is $310, use the equation $0.05(2{,}000) + 0.07x = 310$. Multiply; $100 + 0.7x = 310$. Subtract 100 from both sides; $100 - 100 + 0.7x = 310 - 100$. Simplify; $0.07x = 210$. Divide both sides by 0.07; $\frac{0.07x}{0.07} = \frac{210}{0.07}$. Therefore, $x = \$3{,}000$, which is the amount invested at 7% interest.

**9. c.** Suppose that the cost for one CD, or $d$, is $15. Then the total cost of 10 CDs is $15 · 10 by multiplying the cost per CD by how many CDs are purchased. Therefore, in terms of $d$, the total cost is $d · 10$ which is equivalent to $10d$. Another way to look at this problem is to write the ratios as a proportion, lining up the word labels to help; $\frac{1 \text{ CD}}{d \text{ dollars}} = \frac{10 \text{ CD's}}{? \text{ dollars}}$. Cross-multiply and solve for the unknown number of dollars; $(? \text{ dollars}) · 1 = 10 • d$. Therefore, the total cost is $10d$.

**10. d.** Let $x$ = the number of hours Jack takes to shovel the driveway by himself. In 1 hour Jack can do $\frac{1}{x}$ of the work and Nancy can do $\frac{1}{10}$ of the work. As an equation this looks like $\frac{1}{x} + \frac{1}{10} = \frac{1}{6}$, where $\frac{1}{6}$ represents what part they can shovel in one hour together. Multiply both sides of the equation by the least common denominator, $30x$, to get an equation of $30 + 3x = 5x$. Subtract $3x$ from both sides of the equation; $30 + 3x - 3x = 5x - 3x$. This simplifies to $30 = 2x$. Divide both sides of the equal sign by 2 to get a solution of 15 hours.

## BENCHMARK QUIZ RESULTS

If you answered 8–10 questions correctly, you have a good understanding of how to translate words into mathematical expressions and equations, in addition to solving various types of word problems. After reading through the lesson and focusing on the type(s) of questions you need to refresh, try the quiz at the end of the chapter to ensure that all of the concepts are clear.

If you answered 4–7 questions correctly, you need to review different strategies for solving word problems and translating expressions and equations. Use the chapter as a foundation for careful review and skill building, and pay attention to the sidebars that highlight those important hints and shortcuts. Work through the quiz at the end of the chapter to assess your progress.

If you answered 1–3 questions correctly, extra help and clarification would certainly benefit you in these areas. First, read the chapter and focus on areas that seem unclear. Maybe it has been a while since you have solved these types of problems, so take the time now to freshen up your skills. Use the quiz at the end of the chapter for that extra practice that you need. In addition, you may want to reference a more in-depth and comprehensive book on word problems, such as LearningExpress's *501 Math Word Problems*.

## JUST IN TIME LESSON—TRANSLATING ALGEBRAIC EXPRESSIONS AND SOLVING WORD PROBLEMS

This lesson will review how to translate words into algebraic expressions and how to solve common types of word problems. Topics include:

- translating sentences into mathematical expressions and equations
- setting up and solving consecutive integer problems
- solving other types of word problems including mixture, ratio, geometry, and work problems

## ▶ Translating Expressions and Equations

Translating sentences and word problems into mathematical expressions and equations is similar to translating between two different languages. The key words are the vocabulary that tells what operations should be done and in the order in which they should be evaluated. Use the following chart to help you with some commonly used mathematical key words.

| + | − | × or • | ÷ | = |
|---|---|---|---|---|
| Sum | Difference | Product | Quotient | Equal to |
| More than | Less than | Times | Divided by | Total |
| Added to | Subtracted from | Multiplied by | | Result |
| Plus | Minus | | | |
| Increased by | Decreased by | | | |
| Greater than | Fewer than | | | |

 RULE BOOK

**When translating key words, the phrases *less than* and *greater than* in an algebraic expression do not translate in the same order as they are written in the sentence. For example, when translating the expression *five less than 12*, the correct expression is 12 − 5 not 5 − 12.**

Here is an example of a problem where knowing the key words is necessary:

Twenty less than five times a number is equal to the product of ten and the number. What is the number?

Let's let $x$ equal the number we are trying to find. Now, translate the sentence piece by piece, and then solve the equation.

<u>Twenty less than five times the number</u>  <u>equals</u>  <u>the product of 10 and $x$</u>
$5x - 20$ = $10x$

| The equation is | $5x - 20 = 10x$ |
|---|---|
| Subtract $5x$ from both sides | $5x - 5x - 20 = 10x - 5x$ |
| Divide both sides by 5 | $\frac{-20}{5} = \frac{5x}{5}$; |
| | $-4 = x$ |

In this particular question it is important to realize that the key words *less than* tell you to subtract **from** the number and the key word *product* reminds you to multiply.

## ▶ Problem Solving with Word Problems

There are a variety of different types of word problems you will encounter on various standardized and other kinds of important tests. To help with these types of problems, always begin first by figuring out what you need to solve for and defining your variable(s) as what is unknown. Then write and solve an equation that matches the question asked. The first type of problem is consecutive integer problems.

## ▶ Defining Consecutive Integers

Consecutive integers are integers listed in numerical order that differ by one. Examples of three consecutive integers are 3, 4, and 5 or –11, –10, and –9. Consecutive *even* integers are numbers like 10, 12, and 14 or –22, –20, and –18. Consecutive *odd* integers are numbers like 7, 9, and 11. Consecutive even integers differ by 2, as do consecutive odd integers. When these terms are used in word problems, use the following rule to help define your variables.

 RULE BOOK

When solving a question using regular consecutive integers use $x$, $x + 1$, $x + 2$, etc. to define your integers. When solving a question using consecutive *odd* or *even* integers use $x$, $x + 2$, $x + 4$, etc. to define your integers.

## ▶ Solving Consecutive Integer Problems

To solve a consecutive integer problem, first define the integers. Then write an equation that matches the sentence in the problem.

*Example:*
What is the greater of two consecutive even integers whose sum is 430?
   Since these are two consecutive *even* integers let $x$ = the lesser even integer and let $x + 2$ = the greater even integer. Since *sum* is a key word for addition, the equation is $(x) + (x + 2) = 430$. Combine like terms on the left side of the equation. $2x + 2 = 430$. Subtract 2 from both sides of the equation; $2x + 2 - 2 = 430 - 2$. This simplifies to $2x = 428$. Divide each side by 2; $\frac{2x}{2} = \frac{428}{2}$; $x = 214$, which is the lesser integer. Thus, the greater odd integer is $x + 2 = 216$.

## SHORTCUT

This type of question can also be answered using *guess and check* as a method. If the problem asks for the *sum* of two integers, start by dividing the sum by 2 to find the integers. If the problem asks for the *product* of two integers, start by taking the square root of the product.

## ▶ Mixture Problems

Mixture questions will present you with two or more different types of "objects" to be mixed together. Some common types of mixture scenarios are combining different amounts of money at different interest rates, different amounts of solutions at different concentrations, and different amounts of food (candy, coffee, etc.) that have different prices per pound. The following is an example of a mixture problem with different types of coffee.

*Example:*
How many pounds of coffee that costs $4.00 per pound needs to be mixed with 10 pounds of coffee that costs $6.40 per pound to create a mixture of coffee that costs $5.50 per pound?

For this type of question, remember that the total amount spent in each case will be the price per pound times the number of pounds in the mixture. Therefore, if you let $x$ = the number of pounds of $4.00 coffee, then $4.00(x)$ is the amount of money spent on $4.00 coffee, $6.40(10)$ is the amount spent on $6.40 coffee, and $5.50(x + 10)$ is the total amount spent. Write an equation that adds the first two amounts and sets it equal to the total amount:

$4.00(x) + 6.40(10) = 5.50(x + 10)$
Multiply through the equation: $4x + 64 = 5.5x + 55.$
Subtract $4x$ from both sides: $4x - 4x + 64 = 5.5x - 4x + 55$
Subtract 55 from both sides: $64 - 55 = 1.5x + 55 - 55$
Divide both sides by 1.5: $\frac{9}{1.5} = \frac{1.5x}{1.5};$
$6 = x$

You need 6 pounds of the $4.00 per pound coffee.

## ▶ Ratio Problems

Questions of this type often seem difficult because they usually contain more variables than actual numbers. However, this type of question can be made simpler if numbers are substituted in for the letters. Proportions can also be used to help in these situations, but be sure to use the same units for both ratios. Here's an example.

*Example:*
If Jed earns a total of *a* dollars in *b* hours, how many dollars will he earn in *c* hours?

Think about the problem with actual numbers. If Jed makes 20 dollars in two hours, he makes 10 dollars per hour. This was calculated by dividing \$20 by 2 hours, in other words, $a \div b$. To calculate how much he will make in *c* hours, take the amount he makes per hour and multiply by *c*. Therefore the expression $(a \div b) \bullet c$ simplifies to $\frac{a}{b} \bullet c = \frac{ac}{b}$.

This problem can also be looked at using a proportion. Write the ratios lining up the labels and then cross-multiply to solve for the unknown. For this problem, the proportion would be $\frac{a\,\text{dollars}}{b\,\text{hours}} = \frac{?\,\text{dollars}}{c\,\text{hours}}$. Cross-multiply to get (? dollars) $\bullet$ $b = a \bullet c$. Divide by *b* to get (? dollars) $= \frac{ac}{b}$.

## ▶ Geometry Problems

The key to solving many of the geometry problems presented on exams is knowing the correct formula to apply. Much of the time the formula necessary is a well-known formula, such as area or perimeter. Here is some information on both of these topics.

 RULE BOOK

**The formula for the area of a parallelogram is *area* = *base* × *height*. The family of parallelograms also includes squares, rhombuses, and rectangles. The area of any triangle is *area* = $\frac{1}{2}$ × *base* × *height*, because a triangle is half of a parallelogram. Keep in mind that in this type of problem *base* and *height* can be interchanged with *length* and *width*.**

### ●●●●GLOSSARY

**PERIMETER** the distance around an object. Perimeter is found by adding the sides of the object together.

The following example combines a commonly known formula with algebra to solve for the dimensions of a geometric shape.

*Example:*
The length of a rectangle is two more than its width. The area of the rectangle is 48 square meters. What is the length of the rectangle?

Always start by defining your unknowns in the problem. Let $w$ = the width of the rectangle. Since the length is two more than the width, let $w + 2$ = the length. To find the area of the rectangle, use the formula *area = length × width*, which for a rectangle is the same as *area = base × height*. Since we know that the area is 48, the formula *area = length × width* becomes $48 = w(w + 2)$. To solve, multiply on the right side of the equal sign to get $48 = w^2 + 2w$. Subtract 48 from both sides to get the equation in standard form; $0 = w^2 + 2w - 48$. Factor the quadratic equation into two binomials; $0 = (w + 8)(w - 6)$. Set each factor equal to zero and solve for $w$; $w + 8 = 0$ or $w - 6 = 0$. Therefore, $w = -8$ or 6. Since we are finding the dimensions of a geometric figure, reject the negative value. The width of the rectangle is 6, and the length is $6 + 2 = 8$.

 RULE BOOK

Be sure to reject any negative value in a geometry problem when solving for a dimension of a figure. You cannot have a negative length, or distance.

## ▶ Work Problems

Work problems often present the scenario of two people working to complete the same job. To solve this particular type of problem, think about how much of the job will be completed in one hour. For example, if someone can complete a job in 5 hours, then $\frac{1}{5}$ of the job is completed in 1 hour. If a person can complete a job in $x$ hours, then $\frac{1}{x}$ of the job is completed in 1 hour. Take a look at the next example to see how this is used to solve *work* problems.

*Example:*
Jason can mow a lawn in 2 hours. Ciera can mow the same lawn in 4 hours. If they work together, how many hours will it take them to mow the same lawn?

Think about how much of the lawn each person completes individually. Since Jason can finish in 2 hours, in one hour he completes $\frac{1}{2}$ of the lawn. Since Ciera can finish in 4 hours, then in one hour she completes $\frac{1}{4}$ of the lawn. If we let $x$ = the time it takes both Jason and Ciera working together, then $\frac{1}{x}$ is the amount of the lawn they finish in one hour working together. Then use the equation $\frac{1}{2} + \frac{1}{4} = \frac{1}{x}$ and solve for $x$.

Multiply each term by the LCD of $4x$:
$$4x\left(\tfrac{1}{2}\right) + 4x\left(\tfrac{1}{4}\right) = 4x\left(\tfrac{1}{x}\right)$$

The equation becomes: $\quad 2x + x = 4$

Combine like terms: $\quad 3x = 4$

Divide each side by 3: $\quad \tfrac{3x}{3} = \tfrac{4}{3}$

Therefore, $\quad x = 1\tfrac{1}{3}$ hours

Since $\tfrac{1}{3}$ of an hour is $\tfrac{1}{3}$ of 60 minutes, which is 20 minutes, the correct answer is 1 hour and 20 minutes.

## EXTRA HELP

**For more information and practice on solving different types of word problems, see *501 Math Word Problems* by LearningExpress.**

## TIPS AND STRATEGIES

Translating expressions and solving word problems *can* be done. Use the following tips to assist you on your way to breaking the word problem barrier.

- Keep in mind the key words for translating words into mathematical expressions.
- Start word problems by defining a variable for the value of what you want to solve for.
- Define regular consecutive integers as $x$, $x + 1$, $x + 2$, etc.
- Define both odd and even consecutive integers as $x$, $x + 2$, $x + 4$, etc.
- Remember that the incorrect answer choices for multiple-choice questions are most likely the result of making common errors. Be aware of these traps.
- In questions that use a unit of measurement (such as meters, pounds, etc.) be careful that any necessary conversions took place and that your answer also has the correct unit.
- When using a calculator on a test, be sure to check the reasonableness of your answer. See if your solution is in the ballpark for the range of numbers that would be appropriate for the particular situation.
- If you are really stuck on a question, don't forget to try *guess and check* as a method, especially when the test is multiple-choice.
- Most word problems are just testing skills from the various areas of algebra and mathematics, in general. Don't forget to use what you know to get out of tough spots with word problems.

## CHAPTER QUIZ

Word problems are often some of the most difficult questions you will be asked to solve, but this type can be mastered through practice. Try these practice problems to check your improvement through translating expressions and solving word problems.

1. Which of the following expressions is equivalent to "The sum of three times a number and four"?
   a. $3x - 4$
   b. $3x + 4$
   c. $3x + 12$
   d. $3x \cdot 4$
   e. $3(x + 4)$

2. Which of the following expressions is equivalent to "Three less than five times a number"?
   a. $5(x - 3)$
   b. $3 - 5x$
   c. $5x + 3$
   d. $5x - 3$
   e. $3x - 5$

3. Which of the following expressions is not equivalent to "$8y - 1$"?
   a. one less than eight times $y$
   b. one less than the sum of 8 and $y$
   c. the difference between $8y$ and 1
   d. the product of eight and $y$, decreased by 1
   e. eight multiplied by $y$ and then decreased by one

4. The product of $-3$ and a number is equal to $-36$. What is the number?
   a. $-12$
   b. $-39$
   c. 12
   d. 33
   e. $-33$

5. Twice a number increased by 10 is equal to 32 less than the number. Find the number.
   a. –42
   b. $\frac{22}{3}$
   c. 22
   d. 7
   e. 42

6. The square of a number added to 9 equals 6 times the number. What is the number?
   a. –3
   b. 9
   c. 81
   d. –9
   e. 3

7. The sum of the square of a number and 10 times the number is –24. What is the smaller possible value of this number?
   a. –6
   b. –4
   c. 4
   d. 14
   e. 6

8. The sum of two consecutive even integers is 94. What are the integers?
   a. 42 and 54
   b. 46 and 48
   c. 46 and 44
   d. 47 and 47
   e. 45 and 49

9. What is the lesser of two consecutive positive integers whose product is 240?
   a. –15
   b. 15
   c. –16
   d. 16
   e. 12

10. What is the greater of two consecutive negative integers whose product is 462?
   a. –21
   b. 231
   c. –22
   d. 21
   e. 22

11. The sum of the squares of two consecutive positive odd integers is 130. What is the value of the smaller integer?
   a. 5
   b. 7
   c. 9
   d. 11
   e. 13

12. Jake invested money in two different accounts, part at 12% interest and the rest at 15% interest. The amount invested at 15% was twice the amount at 12%. How much was invested at 12% if the total interest earned was $1,134?
   a. $2,000
   b. $2,100
   c. $2,700
   d. $4,200
   e. $5,700

13. Martha bought $x$ lbs. of coffee that cost $3 per pound and 15 lbs. of coffee at $2.50 per pound for the company picnic. Find the **total** number of pounds of coffee purchased if the average cost per pound is $2.70?
   a. 10
   b. 12
   c. 15
   d. 25
   e. 40

14. The junior class of a high school bought two different types of candy to sell at a school fundraiser. They purchased 50 lbs. of candy at $2.25 per pound and $x$ lbs. at $1.90 per pound. What is the total number of pounds they bought if the total amount of money spent on candy was $169.50?
    a. 50
    b. 60
    c. 80
    d. 90
    e. 150

15. The manager of a garden store ordered two different kinds of pumpkin seeds for her display. The first type cost $1 per packet and the second type cost $1.25 per packet. She bought 50 packets more of the $1.25 seeds than the $1.00 seeds. How many packets of the second type did she purchase if she spent a total of $512.50?
    a. 50
    b. 150
    c. 200
    d. 250
    e. 450

16. If Tonya fills $x$ containers in $m$ minutes, how many minutes will it take to fill $y$ containers?
    a. $\frac{my}{x}$
    b. $mxy$
    c. $\frac{mx}{y}$
    d. $\frac{xy}{m}$
    e. $\frac{y}{mx}$

17. If $e$ eggs are needed to make $c$ cookies, how many eggs are needed to make $10c$ cookies?
    a. $\frac{10c}{e}$
    b. $\frac{10e}{c}$
    c. $10e$
    d. $10c^2e$
    e. $10ce$

**18.** A car travels $m$ miles in $h$ hours. At that rate, how many miles does it travel in 90 minutes?

a. $3mh$

b. $90mh$

c. $\frac{3h}{2m}$

d. $\frac{3m}{2h}$

e. $\frac{45m}{h}$

**19.** The perimeter of a rectangle is 42 inches. What is the measure of its width if its length is 3 inches greater than its width?

a. 3

b. 3.75

c. 4.5

d. 6

e. 9

**20.** Sally owns a rectangular field that has an area of 1,200 square feet. The length of the field is 2 more than twice the width. What is the width of the field?

a. 24

b. 25

c. 34

d. 50

e. 600

**21.** The perimeter of a parallelogram is 100 cm. The length of the parallelogram is 5 cm more than the width. Find the length of the parallelogram.

a. 10

b. 12.5

c. 22.5

d. 25

e. 27.5

**22.** Heather can remodel a kitchen in 10 hours and Steve can do the same job in 15. If they work together, how many hours will it take them to redo the kitchen?

a. 5

b. 6

c. 12.5

d. 15

e. 25

**23.** Anthony can seal the driveway in 150 minutes and John can seal the same driveway in 120 minutes. How many hours will it take them to seal the driveway if they do it together?
a. 1.1
b. 2
c. 30
d. 67
e. 270

**24.** Mary can plant a garden in 3 hours and Zoe can plant the same garden in 6 hours. If they work together, how many hours will it take them to plant the garden?
a. 2
b. 2.1
c. 4.5
d. 6
e. 7.5

**25.** If Joe and Barry work together they can finish a job in 3 hours. If working alone, it takes Joe 8 hours to finish the job, how many hours will it take Barry to do the job alone?
a. 4
b. 4.8
c. 6
d. 7.5
e. 8

## ANSWERS

Use the following answer explanations to monitor your progress in translating expressions and the basics of solving word problems.

**1. b.** *Sum* is a key word for addition and the translation of "Three times a number" is $3x$. This is telling you to add the two parts of the sentence, $3x$ and 4. The translation put together would be $3x + 4$.

**2. d.** When the key words *less than* appear in the sentence it is saying that you will subtract 3 *from* the next part of the sentence, so it will appear at the end of the expression. "Five times a number" can be represented by $5x$. The correct translation of this is $5x - 3$.

**3. b.** Each of the answer choices would translate to $8y - 1$ except for choice **b**. The word *sum* is a key word for addition, and $8y$ means "8 times $y$." Choice **b** would translate to $(8 + y) - 1$ which is not equal to $8y - 1$.

**4. c.** Let $x$ = a number. Since product is a key word for multiplication, the equation is $-3 \cdot x = -36$. Divide both sides by $-3$; $\frac{-3x}{-3} = \frac{-36}{-3}$. The variable is now alone; $x = 12$.

**5. a.** Let $x$ = a number. Now translate each part of the sentence.

Twice a number increased by 10 is:
$$2x + 10$$
32 less than a number is:            $x - 32$
Set them equal as they are in the sentence:
$$2x + 10 = x - 32$$
Subtract $x$ from both sides of the equation:
$$2x - x + 10 = x - x - 32$$
Simplify:            $x + 10 = -32$
Subtract 10 on both sides of the equation:
$$x + 10 - 10 = -32 - 10$$
The variable is now alone:     $x = -42$

**6. e.** Let $x$ = the number. The statement "The square of a number added to 9 equals 6 times the number" translates to the equation $x^2 + 9 = 6x$. Put the equation in standard form and set it equal to zero; $x^2 - 6x + 9 = 0$. Factor the left side of the equation; $(x - 3)(x - 3) = 0$. Set each factor equal to zero and solve; $x - 3 = 0$ or $x - 3 = 0$; $x = 3$ in either case so the number is 3.

**7. a.** Let $x$ = the number. The statement "The sum of the square of a number and 10 times the number is $-24$" translates to the equation $x^2 + 10x = -24$. Put the equation in standard form and set it equal to zero; $x^2 + 10x + 24 = 0$. Factor the left side of the equation; $(x + 4)(x + 6) = 0$. Set each factor equal to zero and solve; $x + 4 = 0$ or $x + 6 = 0$. So, $x = -4$ or $x = -6$. The value of the number is $-4$ or $-6$, the smaller of which is $-6$.

**8. b.** Two consecutive *even* integers are numbers in order like 4 and 6 or $-30$ and $-32$, which are each 2 numbers apart. Let $x$ = the first consecutive even integer. Let $x + 2$ = the second consecutive even integer. *Sum* is a key word for addition so the equation becomes $(x) + (x + 2) = 94$. Combine like terms on the left side of the equation;

$2x + 2 = 94$. Subtract 2 from both sides of the equation; $2x + 2 - 2 = 94 - 2$. Simplify; $2x = 92$. Divide each side of the equation by 2; $\frac{2x}{2} = \frac{92}{2}$; $x = 46$. Therefore, $x + 2 = 48$. The integers are 46 and 48.

**9. b.** Let $x$ = the lesser integer and let $x + 1$ = the greater integer. Since *product* is a key word for multiplication, the equation is $x(x + 1) = 240$. Multiply using the distributive property on the left side of the equation; $x^2 + x = 240$. Put the equation in standard form and set it equal to zero; $x^2 + x - 240 = 0$. Factor the trinomial; $(x - 15)(x + 16) = 0$. Set each factor equal to zero and solve; $x - 15 = 0$ or $x + 16 = 0$. So, $x = 15$ or $x = -16$. Since you are looking for a positive integer, reject the $x$-value of $-16$. If $x = 15$ then $x + 1 = 16$. Therefore, the lesser positive integer would be 15.

**10. a.** Let $x$ = the lesser integer and let $x + 1$ = the greater integer. Since *product* is a key word for multiplication, the equation is $x(x + 1) = 462$. Multiply using the distributive property on the left side of the equation; $x^2 + x = 462$. Put the equation in standard form and set it equal to zero; $x^2 + x - 462 = 0$. Factor the trinomial; $(x - 21)(x + 22) = 0$. Set each factor equal to zero and solve; $x - 21 = 0$ or $x + 22 = 0$. So, $x = 21$ or $x = -22$. Since you are looking for a negative integer, reject the $x$-value of 21. Therefore, $x = -22$ and $x + 1 = -21$. The greater negative integer is $-21$.

**11. b.** Let $x$ = the lesser odd integer and let $x + 2$ = the greater odd integer. The translation of the sentence "The sum of the squares of two consecutive odd integers is 130" is the equation $x^2 + (x + 2)^2 = 130$. Multiply $(x + 2)^2$ out as $(x + 2)(x + 2)$ using the distributive property; $x^2 + (x^2 + 2x + 2x + 4) = 130$. Combine like terms on the left side of the equation; $2x^2 + 4x + 4 = 130$. Put the equation in standard form and set it equal to zero; $2x^2 + 4x - 126 = 0$. Factor the trinomial completely; $2(x^2 + 2x - 63) = 0$ which in turn becomes $2(x - 7)(x + 9) = 0$. Set each factor equal to zero and solve; $2 \neq 0$ or $x - 7 = 0$ or $x + 9 = 0$; $x = 7$ or $x = -9$. Since you are looking for a positive integer, reject the solution of $x = -9$. Therefore, the smaller integer is 7.

**12. c.** Let $x$ = the amount invested at 12% interest. Let $y$ = the amount invested at 15% interest. Since the amount invested at 15% is twice the amount at 12%, then $y = 2x$. Since the total interest was $1,134, use the equation $0.12x + 0.15y = 1,134$. You have two equations with two variables. Use the second equation $0.12x +$

$0.15y = 1,134$ and substitute $(2x)$ for $y$; $0.12x + 0.15(2x) = 1,134$. Multiply on the left side of the equal sign; $0.12x + 0.3x = 1,134$. Combine like terms; $0.42x = 1,134$. Divide both sides by 0.42; $\frac{0.42x}{0.42} = \frac{1,134}{0.42}$. Therefore, $x = \$2,700$, which is the amount invested at 12% interest.

13. **d.** Let $x$ = the amount of coffee at $3 per pound. Let $y$ = the total amount of coffee purchased. If there are 15 lbs. of coffee at $2.50 per pound, then the total amount of coffee can be expressed as $y = x + 15$. Use the equation $3x + 2.50(15) = 2.70y$ since the total number of pounds, $y$, costs $2.70 per pound. To solve, substitute $y = x + 15$ into $3x + 2.50(15) = 2.70y$; $3x + 2.50(15) = 2.70(x + 15)$. Multiply on the left side and use distributive property on the right side; $3x + 37.50 = 2.70x + 40.50$. Subtract $2.70x$ on both sides; $3x - 2.70x + 37.50 = 2.70x - 2.70x + 40.50$. Subtract 37.50 from both sides; $0.30x + 37.50 - 37.50 = 40.50 - 37.50$. Divide both sides by 0.30; $\frac{0.30x}{0.30} = \frac{3.00}{0.30}$. So, $x = 10$ lbs., which is the amount of coffee that costs $3 per pound. Therefore the total amount of coffee is 10 + 15, which is 25 pounds.

14. **c.** Let $x$ = the amount of candy at $1.90 per pound. Let $y$ = the total number of pounds of candy purchased. If there are also 50 lbs. of candy at $2.25 per pound, then the total amount of candy can be expressed as $y = x + 50$. Use the equation $1.90x + 2.25(50) = \$169.50$ since the total amount money spent was $169.50. Multiply on the left side; $1.90x + 112.50 = 169.50$. Subtract 112.50 from both sides; $1.90x + 112.50 - 112.50 = 169.50 - 112.50$. Divide both sides by 1.90; $\frac{1.90x}{1.90} = \frac{57.00}{1.90}$. So, $x = 30$ lbs., which is the amount of candy that costs $1.90 per pound. Therefore the total amount of candy is 30 + 50, which is 80 pounds.

15. **d.** Let $x$ = the number of packets of pumpkin seeds at $1 per packet. Let $y$ = the number of packets of pumpkin seeds at $1.25 per packet. Since there are 50 more packets of the $1.25 seeds than the $1 seeds, $y = x + 50$. Use the equation $1x + 1.25y = 512.50$ to find the total number of packets of each. By substituting into the second equation you get $1x + 1.25(x + 50) = 512.50$. Multiply on the left side using the distributive property; $1x + 1.25x + 62.50 = 512.50$.

Combine like terms on the left side; $2.25x + 62.50 = 512.50$. Subtract 62.50 from both sides; $2.25x + 62.50 - 62.50 = 512.50 - 62.50$. Divide both sides by 2.25; $\frac{2.25x}{2.25} = \frac{450}{2.25}$. So, $x = 200$ packets, which is the number of packets that cost \$1 each. Therefore, the total number of packets, $y$, at \$1.25 each is $200 + 50$, which is 250 packets.

**16. a.** Write the ratios as a proportion, lining up the word labels to help; $\frac{x \text{ containers}}{m \text{ minutes}} = \frac{y \text{ containers}}{? \text{ minutes}}$. Cross-multiply and solve for the unknown number of minutes; (? minutes) $\bullet$ $x = m$ $\bullet$ $y$. Divide both sides by $x$ to get the solution; (? minutes) $= \frac{my}{x}$.

**17. c.** Write the ratios as a proportion, lining up the word labels to help; $\frac{e \text{ eggs}}{c \text{ cookies}} = \frac{? \text{ eggs}}{10c \text{ cookies}}$. Cross-multiply and solve for the unknown number of eggs; (? eggs) $\bullet$ $c = e$ $\bullet$ $10c$. Divide both sides by $c$ to get the solution; (? eggs) $= \frac{10ce}{c} = 10e$.

**18. d.** Write the ratios as a proportion, lining up the word labels to help. Convert 90 minutes to 1.5 hours so that the units used in both ratios are the same; $\frac{m \text{ miles}}{h \text{ hours}} = \frac{? \text{ miles}}{1.5 \text{ hours}}$. Cross-multiply and solve for the unknown number of miles; (? miles) $\bullet$ $h = m$ $\bullet$ $1.5$. Divide both sides by $h$ to get the solution; (? miles) $= \frac{1.5m}{h}$. Recall that $1.5 = \frac{3}{2}$ so the answer choice is $\frac{3m}{2h}$.

**19. e.** Let $x =$ the width of the rectangle. Let $x + 3 =$ the length of the rectangle, since the length is "3 more than" the width. Since perimeter is the distance around the rectangle, the formula is length + width + length + width, $P = l + w + l + w$, or $P = 2l + 2w$. Substitute the "let" statements for $l$ and $w$, and the perimeter ($P$) of 42 into the formula; $42 = 2(x + 3) + 2(x)$. Use the distributive property on the right side of the equation; $42 = 2x + 6 + 2x$. Combine like terms of the right side of the equation; $42 = 4x + 6$. Subtract 6 from both sides of the equation; $42 - 6 = 4x + 6 - 6$. Simplify; $36 = 4x$. Divide both sides of the equation by 4; $\frac{36}{4} = \frac{4x}{4}$. Therefore, $9 = x$.

**20. a.** Let $w =$ the width of the field and let $2w + 2 =$ the length of the field, since the length is two more than twice the width. Since the area of the field is known, use the formula $area = length \times width$. Substitute to get the equation $1{,}200 = w(2w + 2)$. Use the distributive property on the right side of the equal sign; $1200 = 2w^2 + 2w$.

Subtract 1,200 from both sides to get the equation in standard form; $0 = 2w^2 + 2w - 1,200$. Factor the right side of the equation completely; $0 = 2(w^2 + w - 600)$ which becomes $0 = 2(w + 25)(w - 24)$. Set each factor equal to zero and solve; $2 \neq 0$ or $w + 25 = 0$ or $w - 24 = 0$. The two values of $w$ are $-25$ and $24$. Since you can't have negative length, reject the value of $-25$. The width of the rectangular field is 24 ft.

**21. e.** Let $w$ = the width of the parallelogram and let $l$ = the length of the parallelogram. Since the length is 5 more than the width, then $l = w + 5$. Use the formula for the perimeter of a parallelogram: $2l + 2w = 100$. By substituting the first equation into the second for $l$ results in $2(w + 5) + 2w = 100$. Use the distributive property on the left side of the equation; $2w + 10 + 2w = 100$. Combine like terms on the left side of the equation; $4w + 10 = 100$. Subtract 10 on both sides of the equation; $4w + 10 - 10 = 100 - 10$. Divide both sides of the equation by 4; $\frac{4w}{4} = \frac{90}{4}$. So, $w = 22.5$. Therefore the width is 22.5 cm and the length is $22.5 + 5 = 27.5$ cm.

**22. b.** Let $x$ = the number of hours they can remodel the kitchen if they work together. In 1 hour Heather can do $\frac{1}{10}$ of the work and Steve can do $\frac{1}{15}$ of the work. As an equation this looks like $\frac{1x}{10} + \frac{1x}{15} = 1$, where 1 represents 100% of the work. Multiply both sides of the equation by the least common denominator, 30, to get the equation $3x + 2x = 30$. Combine like terms to get $5x = 30$. Divide both sides by 5; $\frac{5x}{5} = \frac{30}{5}$. This gives a solution of 6 hours.

**23. a.** Let $x$ = the number of hours they need to seal the driveway if they work together. Since the times were given in minutes, divide by 60 to get the unit of hours. In 1 hour Anthony can do $\frac{1}{2.5}$ of the work and John can do $\frac{1}{2}$ of the work. As an equation this looks like $\frac{1x}{2.5} + \frac{1x}{2} = 1$, where 1 represents 100% of the work. Multiply both sides of the equation by the least common denominator, 10, to result in the equation $4x + 5x = 10$. Combine like terms to get $9x = 10$. Divide both sides by 9; $\frac{9x}{9} = \frac{10}{9}$. This gives a solution of 1.1 hours.

**24. a.** Let $x$ = the number of hours it takes for them to plant the garden together. In 1 hour Mary can do $\frac{1}{3}$ of the work and Zoe can do $\frac{1}{6}$ of the work. As an equation this looks like $\frac{1x}{3} + \frac{1x}{6} = 1$, where 1 represents 100% of the work. Multiply both sides of the equation by the least common denominator, 6, to result in the equation of $2x + 1x = 6$. Combine like terms to get $3x = 6$. Divide both sides by 3; $\frac{3x}{3} = \frac{6}{3}$. This gives a solution of 2 hours.

**25. b.** Let $x$ = the number of hours it will take Barry to do the job alone. In 1 hour Joe can do $\frac{1}{8}$ of the work and Barry can do $\frac{1}{x}$ of the work. As an equation this looks like $\frac{1}{8} + \frac{1}{x} = \frac{1}{3}$, where $\frac{1}{3}$ represents the part of the job they can complete in one hour together. Multiply both sides of the equation by the least common denominator,

24$x$, to get the equation $3x + 24 = 8x$. Subtract $3x$ from both sides of the equation; $3x - 3x + 24 = 8x - 3x$. This simplifies to $24 = 5x$. Divide both sides of the equation by 5; $x = 4.8$. Therefore, it would take Barry 4.8 hours to complete the job alone.

# ADDITIONAL RESOURCES

**T**his book has given you focused practice and review of your algebra skills. If you need more practice, these resources offer good
places to find what you need to pass your test.

## BOOKS

*501 Algebra Questions* (New York: LearningExpress, 2002).
*Algebra Success in 20 Minutes a Day* (New York: LearningExpress, 2002).
Bobrow, Jerry. *Algebra 1 (Cliffs Quick Review)* (New York: Wiley, 2001).
Yang, Rong. *A-Plus Notes for Algebra* (Redondo Beach, CA: A-Plus
Notes Learning Center, 2000).

## ONLINE RESOURCES

www.aplusmath.com—This website for math students features a large
selection of worksheets, math games, and flashcards.
www.coolmath.com—Improve your math skills through interactive
games, demonstrations, and study tips.

www.math.com—A comprehensive math site with helpful algebra tutorials, practice exercises, and a "worksheet generator" for extra practice.

www.mathforum.org—Drexel University's math center provides information and practice for math students from elementary through graduate school.

www.mathnotes.com—The practice exercises on this site are an excellent supplement to the exercises in algebra books.